Supramolecular Chemistry

Paul D. Beer

Professor in Chemistry, Inorganic Chemistry
Laboratory, University of Oxford

Philip A. Gale

Royal Society University Research Fellow,
Inorganic Chemistry Laboratory, University of Oxford

David K. Smith

Lecturer in Chemistry,
University of York

Series sponsor: **ZENECA**

ZENECA is a major international company active in four main areas of business:
Pharmaceuticals, Agrochemicals and Seeds, Specialty Chemicals, and Biological Products.

ZENECA's skill and innovative ideas in organic chemistry and bioscience create products
and services which improve the world's health, nutrition, environment, and quality of life.

ZENECA is committed to the support of education in chemistry and chemical engineering.

OXFORD
UNIVERSITY PRESS

OXFORD

UNIVERSITY PRESS

Great Clarendon Street, Oxford OX2 6DP
Oxford University Press is a department of the University of Oxford
and furthers the University's aim of excellence in research, scholarship,
and education by publishing worldwide in

Oxford New York
Athens Auckland Bangkok Bogotá Buenos Aires Calcutta
Cape Town Chennai Dar es Salaam Delhi Florence Hong Kong Istanbul
Karachi Kuala Lumpur Madrid Melbourne Mexico City Mumbai
Nairobi Paris São Paolo Singapore Taipei Tokyo Toronto Warsaw

with associated companies in Berlin Ibadan

Oxford is a registered trade mark of Oxford University Press
in the UK and in certain other countries

Published in the United States
by Oxford University Press Inc., New York

A catalogue record for this book is available from the British Library

Library of Congress Cataloging in Publication Data
Data available
ISBN 0 19 850447 0 (Pbk)

Typeset by the author

Printed in Great Britain
on acid-free paper by
The Bath Press, Avon

Series Editor's Foreword

Supramolecular chemistry is one of the most important innovations in inorganic, organic and bio-chemistry in the recent past. The techniques of using inter-molecular forces to assemble chemical structures and to form receptor sites offer a new and subtle approach to synthesis. In addition, the receptor sites may be the basis of sequestering agents, chemical sensors or molecular electronics and molecular mechanical devices.

Oxford Chemistry Primers are designed to give a concise introduction to all chemistry students by providing the material that would usually form an 8–10 lecture course. As well as providing up-to-date information, this series expresses the explanations and rationales that form the framework of current understanding of inorganic chemistry. Paul Beer, Phil Gale and David Smith have produced the first accessible text for undergraduates on this area. It will therefore be of great appeal to other academic staff wishing to design courses in the area, as well as for Third or Fourth year students following such a course.

John Evans
Department of Chemistry,
University of Southampton

Preface

Writing this book has been both a challenge and a pleasure. A challenge because supramolecular science is an ever-expanding area of modern chemistry, making the selection of material with which to introduce this topic to an undergraduate or graduate reader, an endless stream of difficult decisions. A pleasure because above all, supramolecular chemistry is an intellectually stimulating and beautiful area of chemistry in which to work. Many of the systems discussed in this book illustrate great artistry and creativity, and it was refreshing to reconsider them all.

This text is based loosely on a series of final year lectures in Oxford, and provides the basis for such a course to be taught to final year undergraduates or graduates at the start of their research years. Supramolecular chemistry is ideal for the advanced level student who, armed with basic principles from across the whole of modern chemistry, can then apply them to understanding this truly interdisciplinary field.

We are particularly indebted to Prof. R. J. P. Williams for his insights and thoughts on the early stages of each of the chapters included in this primer.

We would also like to thank Prof. Jonathan Sessler (Austin), Dr. Mark Ogden (Curtin University, Perth, WA), Dr. Lance Twyman (Oxford), Dr. Adrian Bisson (Zeneca Specialties) and Mr. Christopher Mellor (Oxford) for reading various incarnations of this primer and pointing out our errors! Additionally, Prof. François Diederich, Anja Schwögler and all David Smith's co-workers in the 'Blacklab' (at ETH, Zürich) are thanked for their great patience during the preparation of this project.

Oxford and Zurich
September 1998

Paul Beer, Philip Gale and David Smith

Contents

1 Introduction

For many years, chemists have synthesized molecules and investigated their physical and chemical properties. The field of *supramolecular chemistry*, however, has been defined as 'chemistry beyond the molecule', and involves investigating new molecular systems in which the most important feature is that the components are held together reversibly by *intermolecular forces, not by covalent bonds*. Chemists working in this area can be thought of as architects combining individual covalently bonded molecular building blocks, designed to be held together by intermolecular forces (supramolecular glue), in order to create functional architectures (Fig. 1.1).

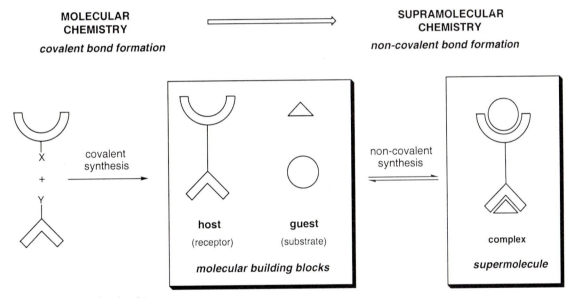

Fig. 1.1 Supramolecular chemistry.

Supramolecular chemistry is a *multidisciplinary* field, and therefore requires a grasp of a range of basic principles. This introduction describes a generalized approach to supramolecular science and provides an indication of the wide ranging interests of chemists working in this area. Biological systems often provide inspiration, organic and inorganic chemistry are required for the synthesis of the pre-designed supramolecular components, and physical chemistry is used to fully understand their properties. Finally, a degree of technical expertise can lead to functioning devices ready for application to the real world. Perhaps the most important assets of a supramolecular chemist, however, are imagination and creativity, which have given rise to a wide range of beautiful and functional systems, as the remaining chapters of this Primer will illustrate using specific examples.

As supramolecular chemistry is a multidisciplinary field, the terminology used has evolved from various sources including inorganic, organic, physical and biological chemistry. This has lead to a number of different terms being used to describe the host and guest. This can be quite confusing if you are new to this area of chemistry. For example, a receptor molecule (an enzyme in biological chemistry) might also be referred to as a host molecule or a binding agent. The substrate that the molecule binds might also be referred to as a guest species.

Synonymous terminology used:

host	guest
ligand	metal
enzyme	substrate
receptor	substrate
receptor	drug
antibody	antigen

1.1 Complementarity in biology: how things fit together

One of the most important concepts introduced in this Primer is *complementarity*. The most elegant examples of complementary molecular systems are biological in origin and so strictly outside the scope of this book. They do, however, provide a deep source of inspiration and lay down a challenging standard for supramolecular chemists to attempt to match.

An enzyme may catalyze a single reaction with high or total specificity; the active site of the enzyme being complementary to the substrate (or more specifically to the transition state of the catalysed reaction). In other words, the size, shape and position of the binding sites within the active site are ideal for specific substrate recognition. Emil Fischer described this idea in 1894 as the *'Lock and Key' principle*. Figure 1.2 shows a stylized representation of the lock and key. The arrangement of binding sites in the host (lock) is complementary to the guest (key) both sterically and electronically.

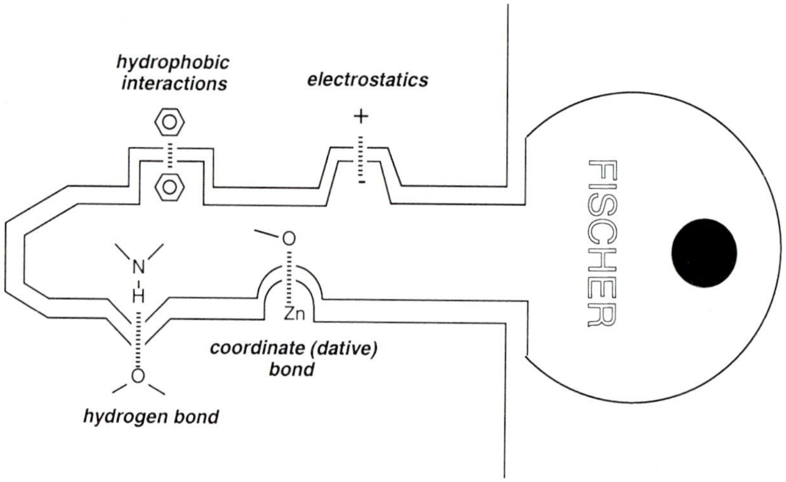

Fig. 1.2 The lock and key principle: receptor sites in the host (lock) are complementary to the guest (key) (see Section 1.2 for a discussion of these forces in more detail).

An example of the lock and key principle in Nature is provided by carboxypeptidase-A, an enzyme that selectively catalyses the hydrolysis of the C-terminal amino-acid residues of proteins. It is the zinc ion at the enzyme active site that catalyses the removal of an amino-acid residue from the carbon end of the polypeptide chain. Figure 1.3 shows an illustration of the active site of carboxypeptidase-A with a bound peptide chain. The anionic carboxylate terminus of this bound peptide is hydrogen bonded to the positively charged guanidinium group of Arg-145 forming a salt-bridge (see Section 3.6).

The active site also possesses a hydrophobic pocket which can accommodate the side chain of the substrate's C-terminal residue. Therefore, if the C-terminal of the polypeptide chain possesses a hydrophobic residue (e.g. phenylalanine), then the enzyme is more likely to bind that particular peptide chain than one lacking such a group. This hydrophobic pocket provides the enzyme with a mechanism by which it can select *specific peptides* for hydrolysis, accomplished by the Lewis acidic zinc centre.

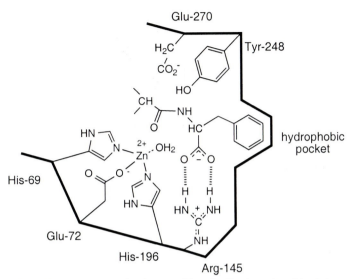

Fig. 1.3 The active site of carboxypeptidase-A with a bound peptide chain.

Perhaps the most important example of complementarity, however, is base pairing in the *DNA double helix*. A single DNA strand is composed of purine and pyrimidine bases linked to a backbone of phosphorylated sugars. The DNA double helix combines two antiparallel strands which are held together by complementary hydrogen bonds between pairs of bases.

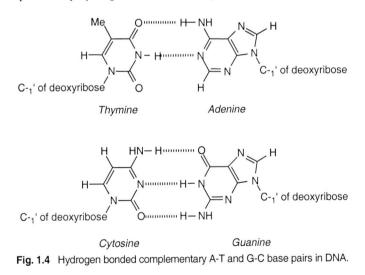

Fig. 1.4 Hydrogen bonded complementary A-T and G-C base pairs in DNA.

The hydrogen bond donors and acceptors on the nucleic acid bases are arranged such that adenine (A) forms two hydrogen bonds with thymine (T), whereas guanine (G) complements cytosine (C) with the formation of three hydrogen bonds, therefore only A–T and G–C base pairs usually form (Fig.

1.4). This complementarity extends from individual base pairs to the whole double helical DNA structure in which one strand (e.g. CCTTATAGAGG) is complementary to the other (e.g. GGAATATCTCC).

Valinomycin is a naturally occurring macrocyclic antibiotic, first isolated in 1955, that selectively transports potassium cations across mitochondrial membranes in the presence of sodium cations. It has a cyclic structure consisting of three identical repeated fragments each containing D-hydroxyisovaleric acid, D-valine, L-lactic acid and L-valine (Fig. 1.5).

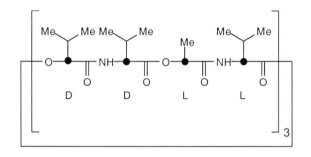

Fig. 1.5 Valinomycin.

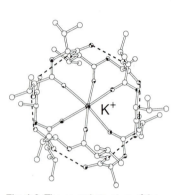

Fig. 1.6 The crystal structure of the valinomycin-K^+ complex.

K^+ is complexed by the electronegative oxygen atoms of the antibiotic ester groups and once it is encapsulated within the macrocycle it can be efficiently transported through the hydrophobic membrane. The membrane is lipophilic and so will not normally allow charged species to pass through. The exterior of the K^+–valinomycin complex, however, has peripheral alkyl groups that are 'greasy' providing the complex with solubility in the membrane and allowing transport of the metal ion. The lipophilic exterior can be seen in the crystal structure of the potassium complex (Fig. 1.6). The conformation adopted by valinomycin on K^+ complexation is stabilized by six NCO···HN hydrogen bonds around the periphery of the macrocycle.

1.2 Non-covalent interactions: the forces at our disposal

This book, however, is concerned with the design, synthesis, coordination and assembly properties of molecules *designed by the chemist* — not by Nature. The glue used by supramolecular chemists to hold molecules together is non-covalent, and there are a number of such interactions that can be utilized. They include:

(a)   electrostatics (ion–ion, ion–dipole and dipole–dipole);
(b)   hydrogen bonding;
(c)   π–π stacking interactions;
(d)   dispersion and induction forces (van der Waals forces);
(e)   hydrophobic or solvatophobic effects.

The bond energy of a typical single covalent bond is around 350 kJmol^{-1} rising up to 942 kJmol^{-1} for the very stable triple bond in N_2. The strengths of many of the non-covalent interactions used by supramolecular chemists are generally much weaker ranging from 2 kJmol^{-1} for dispersion forces, through to 20 kJmol^{-1} for a hydrogen bond to 250 kJmol^{-1} for an ion–ion interaction. The power of supramolecular chemistry lies in the combination of a number of weak interactions, allowing strong and selective recognition of specific guests to be achieved.

• *Electrostatic interactions* (such as the ion–dipole interactions that operate in valinomycin) are based on the Coulombic attraction between opposite charges (Fig. 1.7). Ion–ion interactions are non-directional, whilst for ion–dipole interactions the dipole must be suitably aligned for optimal binding efficiency. The high strength of electrostatic interactions has made them a prized tool amongst supramolecular chemists for achieving strong binding. There are many receptors for cations (crown ethers, cryptands and spherands discussed in Chapter 2) and anions (protonated or alkylated polyammonium macrobicycles discussed in Chapter 3) which employ electrostatic interactions to hold the guest in place.

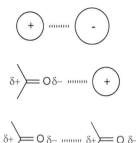

Fig. 1.7 Electrostatic interactions (ion–ion, ion–dipole, dipole–dipole).

• Arrays of *hydrogen bonds*, such as those employed in biological systems (e.g. the DNA double helix), have been utilized in receptors designed to coordinate neutral organic species such as barbiturates, short chain alcohols and amides (Chapter 5), and also anions (Chapter 3). The directional nature of hydrogen bonds, combined with the precision with which the individual components can be built into molecular systems has made them especially attractive to molecular designers. This has facilitated the construction of complex architectures (such as rosettes see Chapter 6).

• *π–π stacking forces* occur between systems containing aromatic rings (Fig. 1.8). Attractive interactions can occur in either a 'face-to-face' or 'edge-to-face' manner (for example benzene crystallizes in a 'herring-bone' arrangement maximising edge-to-face contacts). Current theories suggest this attractive force is electrostatic in nature. Some very elegant receptors have been synthesized employing π–π interactions, including a receptor for benzoquinone (Chapter 5).

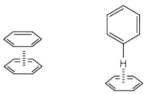

face-to-face   edge-to-face

Fig. 1.8 π–π stacking interactions.

• *Dispersion forces* (or induced dipole-induced dipole interactions) are attractive forces between molecules that occur when instantaneous dipoles in the electron clouds around each molecule interact favourably. These van der Waals forces are believed to provide additional enthalpic stabilisation to the coordination of a hydrophobic guest into a hydrophobic cavity. They are, however, of a very general nature and so it is difficult to design receptors specifically to take full advantage of them. One such system may be a self-assembled 'tennis-ball' that can encapsulate xenon atoms (Chapter 6).

• The *hydrophobic effect* (Fig. 1.9) is the specific driving force for the association of apolar binding partners in aqueous solution. Water molecules around the apolar surfaces of a hydrophobic cavity arrange themselves to form a structured array. Upon guest complexation the water molecules are released and become disordered. This results in a favourable increase in entropy. In addition, there is believed to be an enthalpic component to the hydrophobic effect. The hydrogen bonds between water molecules are larger than the interactions between the water molecules and apolar solutes, providing an enthalpic force for apolar guest coordination (i.e. when the water in the apolar cavity is released into the bulk solvent it can maximize its hydrogen bonding

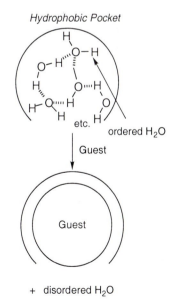

Fig. 1.9 Origin of the hydrophobic effect.

interactions). Receptors containing hydrophobic interior cavities designed to encapsulate organic guest molecules in aqueous solution include the cyclophanes and cyclodextrins (Chapter 5).

• *Classical coordination chemistry* (i.e. the coordination of metals by ligands donating two electrons to form a dative bond) although *not strictly* a non-covalent interaction is also widely used in supramolecular chemistry. The geometric requirements of metal ions, combined with the design of specific ligands has permitted the construction of complex and eye-catching molecular topologies including catenanes, double and triple helices, and molecular grids (Chapter 6).

• *Steric repulsion.* This diminishes the strength of interactions as two molecules cannot occupy the same space. As may be expected from the lock and key analogy, however, it can play a very important role in determining the selectivity of a receptor species for a particular substrate and the stability of a specific complex.

The forces described above can be used individually, but it is more usual for the supramolecular chemist to use these forces in concert to maximize the *selectivity* and *tunability* of the new receptor and also increase the strength of the complex formed.

1.3 Design principles: chelate and macrocyclic effects

Any architect has certain guiding principles to help design successful structures. The situation is no different for the molecular architect who must employ the non-covalent forces described above in order to recognize and bind a specific guest. *Design principles* are therefore applied in order to achieve the desired intermolecular interaction, with a number of factors being used to increase the strength of the intended host-guest complex. In particular, chelating or macrocyclic ligands are frequently employed due to the high thermodynamic stability of their complexes.

The *chelate effect* refers to the enhanced stability of a complex containing chelate rings as compared to a similar system containing fewer or no rings. This is most clearly illustrated by comparing two different ligands: ethylene diamine and ammonia (Fig. 1.10).

The metal complex containing *bidentate* ethylene diamine (1,2-diaminoethane) is almost ten orders of magnitude more stable than that containing no chelating ligands. The reason for this lies in thermodynamic considerations. An increase in the overall binding constant (β) corresponds to a more negative value of $\Delta G°$ (for a precise definition of β see Fig. 1.11). This would result from either a more negative enthalpy or a more positive entropy on complexation. For ammonia binding, six ligands replace six

$$[Ni(H_2O)_6]^{2+}{}_{(aq)} + 6NH_{3(aq)} \rightleftharpoons [Ni(NH_3)_6]^{2+}{}_{(aq)} + 6H_2O_{(aq)} \quad \log \beta = 8.61$$

$$[Ni(H_2O)_6]^{2+}{}_{(aq)} + 3en_{(aq)} \rightleftharpoons [Ni\ en_3]^{2+}{}_{(aq)} + 6H_2O_{(aq)} \quad \log \beta = 18.28$$

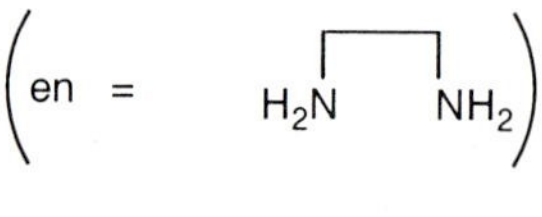

$$\left(en\ =\ H_2N \overline{\qquad} NH_2 \right)$$

Fig. 1.10 Chelating ligands bind metal ions more strongly.

waters and so the number of independent species in solution remains the same. Ethylene diamine, however, is bidentate so three ligands displace six waters, increasing the number of independent species in the system and causing an increase in entropy, thus lowering $\Delta G°$. Enthalpic factors also play a role. The polar amino groups, which are separated in ammonia, are covalently brought together in ethylene diamine overcoming a part of their mutual repulsion, making coordination energetically more favourable. Another more subtle factor is the increase in basicity (and consequently metal binding ability) of the ethylene diamine amino groups resulting from the inductive effect of the alkyl bridge.

$$\Delta G° = -RT \ln \beta \quad (1.1)$$

$$\Delta G° = \Delta H° - T\Delta S° \quad (1.2)$$

Stepwise Binding Constants:

$$H + G \rightleftharpoons HG \qquad K_1 = \frac{[HG]}{[H][G]}$$
(Host) (Guest)

$$HG + G \rightleftharpoons HG_2 \qquad K_2 = \frac{[HG_2]}{[HG][G]}$$

$$HG_2 + G \rightleftharpoons HG_3 \qquad K_3 = \frac{[HG_3]}{[HG_2][G]}$$

$$HG_{n-1} + G \rightleftharpoons HG_n \qquad K_n = \frac{[HG_n]}{[HG_{n-1}][G]}$$

Overall Binding Constants:

$$H + G \rightleftharpoons HG \qquad \beta_1 = \frac{[HG]}{[H][G]}$$

$$H + 2G \rightleftharpoons HG_2 \qquad \beta_2 = \frac{[HG_2]}{[H][G]^2}$$

$$H + 3G \rightleftharpoons HG_3 \qquad \beta_3 = \frac{[HG_3]}{[H][G]^3}$$

$$H + nG \rightleftharpoons HG_n \qquad \beta_n = \frac{[HG_n]}{[H][G]^n}$$

Therefore:

$$\beta_n = \prod_1^n K_n \qquad [\text{e.g. } \beta_3 = K_1 \times K_2 \times K_3]$$

Fig. 1.11 Definition of stepwise (K) and overall (β) binding constants.

Interestingly, from five membered chelate rings upwards, the chelate effect decreases in magnitude with increasing ring size (four membered rings are unstable due to extreme ring strain). This can be explained by considering the configurational entropy of the chelate chain. The longer the chain, the higher the configurational entropy and so ring formation becomes increasingly improbable.

It is perhaps surprising that five membered chelate rings are considerably more stable with large metal ions than six membered chelate ring systems. This can be explained by a consideration of ring strain. The chair form of cyclohexane is shown in Fig. 1.12. In an unstrained six membered chelate ring containing sp^3 hybridized atoms the bond angles will be the same as those in the chair form of cyclohexane, namely 109.5°. Therefore in an unstrained six membered chelate ring system (we could imagine replacing one carbon in cyclohexane with a metal and the two adjacent carbon atoms with

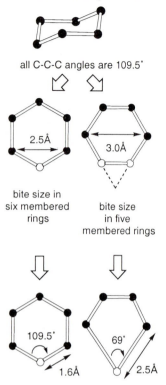

Chair form of cyclohexane

all C-C-C angles are 109.5°

2.5Å

3.0Å

bite size in six membered rings

bite size in five membered rings

109.5°

69°

1.6Å

2.5Å

Fig. 1.12 Complexes containing five membered chelate rings (bite size = 3.0Å) are more stable with larger metals than those containing six membered rings (bite size = 2.5Å).

Table 1.1 Thermodynamic parameters for the formation of zinc complexes with macrocyclic and acyclic aza ligands at 25°C.

	1.1	1.2
log K	11.25	15.34
$-\Delta H^\circ$ (kJmol^{-1})	44.4	61.9
ΔS° (JK^{-1}mol^{-1})	66.5	85.8

nitrogens) the bite size is only 2.5Å. However a five membered chelate ring system (we could imagine removing two carbon atoms and replacing them with a single metal such that the C-C-M bond angles in the ring are 109.5°, then replace the carbon atoms adjacent to the metal with nitrogen atoms) has a larger bite size of 3.0Å and has more space (2.5Å compared to 1.6Å) to incorporate a metal ion guest without straining the chelate bridge (Fig. 1.12). So five membered chelate rings containing the larger Pb^{2+} are more stable than five membered chelate rings containing the smaller Cu^{2+}.

The *macrocyclic effect* is related to the chelate effect and refers to the increased thermodynamic stability of macrocyclic systems compared to their acyclic analogues. Consider the complexation of zinc (II) by two ligands: one acyclic (**1.1**) and one macrocyclic (**1.2**) (Fig. 1.13).

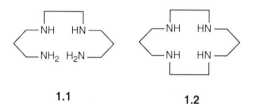

1.1

1.2

Fig. 1.13 Acyclic and macrocyclic aza-ligands.

The formation of zinc (II) complexes by **1.1** and **1.2** have the thermodynamic parameters shown in Table 1.1. In this instance the macrocyclic effect causes an increase in complex stability by just over four orders of magnitude. This increased stability is caused by a combination of enthalpic and entropic factors, with the relative importance of each term varying from case to case, although enthalpy frequently provides a dominant contribution. Macrocyclic hosts are less heavily solvated than their acyclic analogues and therefore less energy is required for desolvation (coordination is more enthalpically favourable). Macrocyclic ligands are also less flexible and consequently have less disorder to lose on complexation than their acyclic analogues (in other words, coordination is more entropically favourable because of the relative rigidity of the receptor; see Sections 2.3, 2.4 and 2.5). Additionally, macrocyclic complexes are also more kinetically inert than their acyclic analogues (i.e. their rates of complexation and decomplexation are generally slower).

The enhanced binding of guest species provided by chelating or macrocyclic hosts has been employed in the design of many receptors operating through a variety of intermolecular forces. *Molecular design* is a key theme of this book, and the examples discussed in later chapters are chosen to highlight these and other design principles which supramolecular chemists frequently utilize.

1.4 Characterising supramolecular systems

The supramolecular chemist therefore acts as an architect, controlling a range of intermolecular interactions in order to achieve strong and specific recognition. Once the design and synthesis have been achieved, however, the supramolecular system must be adequately *characterized*. In particular, the following information about the supermolecule is important:

* what is the structure, is it as planned?
* how rapidly is it formed (kinetics)?
* how strong are the interactions (thermodynamics)?

The accurate characterisation of supramolecular complexes is a large and very complex field and the scope of this book only allows an indication as to how this goal is achieved. In particular, the use of nmr, often the first point of reference, will be highlighted.

Structural Information

Crystallography: Perhaps the most convincing evidence of a supramolecular interaction is a crystal structure of the host-guest complex (e.g. Fig. 1.14, see also Section 2.3). Crystallography clearly shows the binding site, and if the complex is as planned by the molecular designer. It also gives information about the interactions that hold the guest in place. A crystal structure, however, is only valid for the solid state, as factors such as crystal packing (which do not come into play in solution or the gas phase) may alter the properties of the supermolecule. Supramolecular chemists are often interested in the solution phase, and to understand this, alternative methods are used.

Nmr: Many atomic nuclei (e.g. ^{1}H, ^{13}C, ^{31}P) possess nuclear spin and are thus detectable by nuclear magnetic resonance (nmr). The frequency at which a particular nucleus resonates is dependent on the electronic environment in which it finds itself. It is therefore a sensitive technique for monitoring the interactions between molecules. A typical experiment is called an *nmr titration*. The nmr spectrum of a solution of the host in deuterated solvent is measured, and then the guest is added to this solution in small aliquots. The nmr resonances of the host are monitored, and if binding occurs, the guest perturbs their electronic environments. Protons which form specific hydrogen bonds to the guest, or are located in parts of the receptor with close guest contact are most strongly affected. Structural information about the supermolecule can therefore be obtained. This technique is not limited to ^{1}H nmr, and titration experiments may be performed by monitoring any nmr active nucleus in the host or the guest provided that the electronic environment of the nucleus is perturbed on binding. Nmr is often used to prove the stoichiometry of binding, which is obtained from the *method of continuous variation* (Fig. 1.15).

Nuclear Overhauser Effect Spectroscopy (NOESY) is also of increasing importance. NOESY provides an nmr spectrum correlating resonances that are physically close together. The geometry of coordination between host and guest can often be exactly defined. This is particularly useful for more complex assembled systems such as helices held together by intermolecular forces, which have non-trivial spectra.

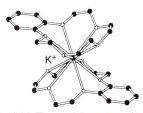

Fig. 1.14 The crystal structure of two benzo-15-crown-5 molecules forming a 'sandwich complex' with a potassium cation (see section 2.3).

Method of continuous variation: A series of solutions containing [Host]+[Guest]=Constant is made and the solution properties monitored in order to follow [Complex] against the [Host]/([Host]+[Guest]) ratio. When the value of [Complex] reaches a maximum then the value of {Host}:[Guest], which is obtained from the [Host]/([Host]+[Guest]) ratio, corresponds to the complex stoichiometry. This can be clearly seen graphically, the stoichiometry being read off the graph, which is referred to as a Job plot (Fig. 1.14).

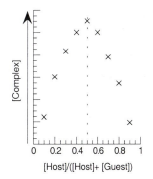

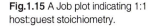

Fig.1.15 A Job plot indicating 1:1 host:guest stoichiometry.

Electrospray mass spectrometry has often been used for characterisation, as it provides a mild solution phase method of ionisation. A solution of the complex being investigated is sprayed as a mist into a strong electrostatic field, which ionizes the 'supermolecule'. The solvent then evaporates from the tiny droplets until the charged complexes repel one another and fly into the gas phase, where they can be sampled by the mass spectrometer.

Other Techniques: Other receptor properties can be monitored in order to study the effect of guest molecules. *UV–visible spectroscopy* is especially effective for investigating π-electron systems or transition metals, as their spectra can be strongly perturbed on complex formation. Increasingly, *mass spectrometry* is being used to ascertain the mass of host-guest complexes, but the ionisation method must be mild, otherwise the complex will be broken into its constituent pieces, rather than flying through the spectrometer as a discrete unit. As the range of supramolecular chemistry expands, and the complexes become ever larger (see Chapter 6), techniques such as chromatography using *size exclusion gel*, on which the largest molecules migrate the most rapidly, become increasingly important for finding the aggregate mass.

Kinetic Information

As explained above, the host's nmr resonances are perturbed by the addition of guest. Complexation is a dynamic exchange between the bound and unbound forms of the host, and the nmr response therefore provides an insight into the binding kinetics.

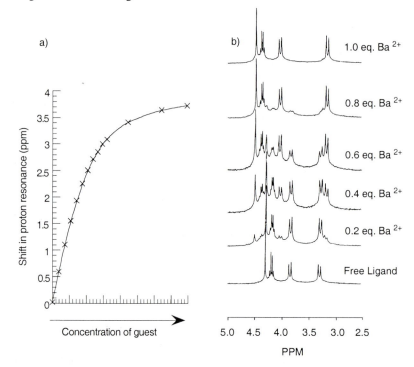

Fig. 1.16 (a) A titration curve from an nmr experiment where guest binding is kinetically fast leading to a time averaged spectrum and shifts in proton resonances (left) and (b) a series of nmr spectra in which barium binding is kinetically slow leading to resonances corresponding to both receptor and complex being visible (right).

If the binding is *kinetically fast* compared to the frequency separation of free and complexed host nmr resonances (i.e. complex lifetime $< 10^{-2}$-10^{-3}s), then the host's nmr resonances are observed as an average peak. On the addition of increasing quantities of guest, this time averaged peak shifts continuously until the receptor is saturated. A titration curve having a distinctive shape can then be extracted (Fig. 1.16a). Initially, the host is strongly perturbed by the addition of small amounts of guest, but at higher concentrations, it becomes saturated by the guest, and is not further perturbed.

If, however, binding is *kinetically slow* compared to the nmr timescale, then a time averaged nmr peak is not observed on guest addition. Instead, the resonances for free host gradually diminish in intensity, whilst resonances for the host-guest complex (at different chemical shifts) grow (Fig. 1.16b).

Quantitative information can also be obtained. For any exchange process, such as complexation, the kinetics can be varied by changing the temperature. If a solution exhibiting slow exchange is heated, the two individual peaks gradually broaden and merge. Eventually, a point intermediate between fast and slow exchange is reached at which a single broad peak is observed. This is referred to as *coalescence* and the rate of complexation at the coalescence temperature can be derived (eqn. 1.3 – only applies to 1:1 singlets).

$$k(ct) = \frac{\pi \delta v}{\sqrt{2}} \quad (1.3)$$

k(ct): rate of complexation at coalescence temperature
δv: splitting between the nmr peaks under the condition of slow exchange

Other techniques for investigating host-guest interactions operate on different timescales. In UV-Vis spectroscopy, for example the lifetime is typically 10^{-15}s, and as this rate is faster than a diffusion-controlled process, all recognition events investigated by this method exhibit slow exchange.

Thermodynamic Information

As described on page 6, the *binding constant* is related to the free energy of binding, ΔG (eqn. 1.1) (Fig. 1.11 defines stepwise (K) and overall (β) binding constants). This free energy is in turn related to the entropy (ΔS) and enthalpy (ΔH) of binding (eqn. 1.2). Many binding processes are enthalpically favourable (energetically favourable interactions form between host and guest) but entropically unfavourable (loss of freedom on coordination of the guest).

$$\Delta G = \Delta H - T\Delta S \quad (1.2)$$

$$\frac{d\Delta G}{dT} = -\Delta S \quad (1.4)$$

$$\Delta G = -RT \ln K \quad (1.1)$$

$$\ln K = \frac{\Delta S}{R} - \frac{\Delta H}{RT}$$

$$\frac{d \ln K}{dT} = \frac{\Delta H}{RT^2} \quad (1.5)$$

Titration experiments can provide thermodynamic data. Nmr experiments will, once again, be taken as an example. If the binding is kinetically slow, the relative concentrations of host and complex can be obtained from integration of the proton resonances. For a 1:1 complex, this allows direct calculation of the binding constant K_1, as all three concentrations (host, guest and complex) are known. If the binding is kinetically fast, the titration curve (Fig. 1.16a) contains all the necessary information. The 'sharpness' of curve reflects the affinity of the host for the guest. *Computer programs* are routinely used to find the binding constant from this data by using a non-linear least squares procedure to fit a theoretical model of the complexation process to the experimental data. Performing titration experiments at different temperatures yields different binding constants, and this data can be used to yield values for ΔH and ΔS (assuming they remain invariant with temperature) (see eqns. 1.4 and 1.5).

Another method for determining thermodynamic parameters is calorimetry, in which the quantity of heat evolved or absorbed on complex formation is measured. Once again, a titration is performed, with the temperature of a solution of host being measured as a function of the quantity of guest added.

$$E = E^{\ominus} + \frac{RT}{nF} \ln a \qquad (1.6)$$

E: electrode potential
a: activity of observed entity
R: gas constant
n: no. of electrons transferred
F: Faraday constant

Obviously, the heat change on complexation is directly linked to the desired thermodynamic values.

Potentiometric methods have been particularly important in aqueous solution for the monitoring the binding constants of ionic guests. The Nernst equation (eqn. 1.6) relates the potential of a solution to the activity of a particular ion, which can be approximated by its concentration. If, due to complexation, this concentration varies, then so will the potential of the solution. This potential can be measured by an ion selective electrode (e.g. a pH electrode for protons), and the response of this electrode will then be directly dependent on the degree of complexation of the guest ion. Once again, a titration is performed by addition of guest ions to the host, with the solution potential being monitored. The potentiometric data generated can then be analysed by computer, to yield binding constants. Potentiometry is particularly powerful as it allows the routine thermodynamic treatment of complicated multi-step complexation processes in aqueous solution.

There are many additional approaches to the thermodynamic characterisation of supramolecular aggregates including solubility measurements, extraction experiments etc., but an in depth treatment lies beyond the scope of this book.

1.5 Solvent Effects

So far, the focus has been on the individual components designed to perform specific recognition as if they existed in a vacuum. The *solvent*, however, is often far from inert, playing an *important role* in recognition processes (Fig. 1.17). If the solvent strongly solvates either host, guest or complex, or if the solvent interacts strongly with itself, then this can have dramatic effects on the host-guest equilibrium. Desolvation possesses an unfavourable enthalpy (energy is required to break bonds to solvent molecules) and a favourable entropy (increase in disorder on release of solvent). If ignored, the effect of solvent can undermine much of the careful receptor design, making recognition non-specific and inefficient.

HOST.Sol + GUEST.Sol ⇌ HOST–GUEST.Sol + Sol.Sol

Fig. 1.17 Host–Guest equilibrium including solvent.

a) The *hydrophobic effect* (Section 1.2) provides an example of the way in which solvent can influence molecular recognition. As discussed above, this source of binding energy is a combination of entropic liberation of water and enthalpically favourable solvent–solvent hydrogen bonding interactions. In addition, water can only solvate the large apolar surfaces of the uncomplexed host and guest very poorly. The factors pull the complexation equilibrium (Fig. 1.17) to the right hand side.

b) For *electrostatic interactions*, the *dielectric constant* (Table 1.2) of the solvent plays an important role in controlling binding strength. The dielectric constant of a solvent measures its bulk polarity and reflects the dipole moment of an individual molecule of solvent. Solvents with high dipole moments interact more effectively with charged (or partially charged) species, shielding them from one another and diminishing the strength of their interactions with each other (eqn. 1.7). This is reflected, for example, in the strength of ion pair formation, which is very high in solvents of low polarity, and diminishes as the dielectric constant increases.

$$\text{Energy} = \frac{kq_1q_2}{\varepsilon r} \quad (1.7)$$

k: constant
q_1 and q_2: charges
ε: dielectric constant
r: separation

c) The *donor-acceptor ability* of the solvent also plays a significant role in controlling molecular recognition (Fig. 1.18). There are a number of different empirical measures of a solvent's ability to donate (or accept) an electron pair. One popular scale is provided by the *Gutmann donor and acceptor numbers* (Table 1.2). A good donor solvent, for example, solvates cationic species very efficiently and will consequently compete against any receptor for such guests, weakening their recognition. A good acceptor solvent, meanwhile, solvates anions very effectively and therefore weakens anion binding. Changing the solvent can also alter the selectivity, as one guest may be more strongly solvated than the other.

Donor-acceptor ability is also of great importance in *hydrogen bonding recognition*. A good electron pair donor will readily accept a hydrogen bond, whilst a good electron pair acceptor is usually a good hydrogen bond donor (Fig. 1.19). If the solvent is a good hydrogen bond donor or acceptor, this disrupts and weakens any recognition which depends on hydrogen bonds, as the unbound host and guest are solvated more effectively, pulling the equilibrium in Figure 1.17 to the left hand side. This problem is particularly acute in aqueous solution, as water is both an excellent hydrogen bond donor and acceptor. Solvents which strongly disrupt molecular recognition in this way are referred to as *competitive solvents*.

good donor
and
acceptor

good donor
poor acceptor

Fig. 1.18 Donor and acceptor properties of solvents.

Table 1.2 Gutmann Donor and Acceptor Numbers and dielectric constants for selected solvents (a: donor number for water when measured in bulk solvent).

	Donor number (H-bond acceptor)	Acceptor number (H-bond donor)	Dielectric constant (ε)
H_2O	18.0 (33.0)[a]	54.8	80.1
CH_3SOCH_3	29.8	19.3	46.7
CH_3CN	14.1	18.9	36.6
CH_3OH	19.0	41.5	33.0
CH_3COCH_3	17.0	12.5	21.0
C_4H_8O (THF)	20.0	8.0	7.5
C_6H_{14} (Hexane)	no donor atoms	0.0	1.9

Solvent effects, however, are rarely simple, and often operate in complex combinations. For example, donor-acceptor ability and dielectric effects both play important roles in ion recognition. With a reasonable understanding of the roles solvents can play, however, we are able to design receptors with new selectivities and enhanced binding strengths in a whole range of different solvent media.

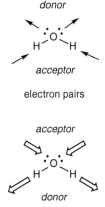

donor

acceptor

electron pairs

acceptor

donor

hydrogen bonds

Fig. 1.19 An electron pair donor is hydrogen bond acceptor, and an electron pair acceptor can be a hydrogen bond donor.

1.6 Informed Design

After considering intermolecular forces, designing and synthesising a receptor, and studying its interactions in solution; what happens next?

Normally, studies of the binding strength and selectivity provide an insight into how the receptor could be improved. In this way, the supramolecular approach feeds back, and a second generation of more effective, selective receptors can be synthesized (Fig. 1.20). Eventually, the properties of the system become so finely tuned that they are suitable for technological application and can be incorporated into a *functional device* (see Chapter 7).

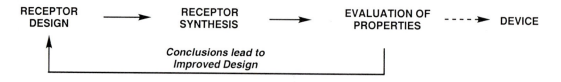

Fig. 1.20 The concept of informed design eventually leads to functional devices.

1.7 Summary and Overview

Supramolecular chemistry concerns the design and synthesis of individual molecular components which are then used to build non-covalently linked molecular assemblies. In this chapter we have introduced the approach to supramolecular chemistry in a general way, showing how biological examples of non-covalently linked molecular systems often provide inspiration that fuels the imagination of the supramolecular chemist.

The remaining chapters of this book focus on specific examples from supramolecular chemistry in order to illustrate important principles of design and molecular recognition. Chapters Two and Three discuss the binding of cations and anions respectively, whilst Chapter Four discusses the recognition of both together in the form of an ion pair or zwitterion. Chapter Five discusses the binding of neutral molecules. Chapter Six aims to bring together themes from the previous Chapters and shows how supramolecular chemistry can build complex, spectacular architectures held together with a variety of intermolecular forces. Finally, Chapter Seven will present current applications of Supramolecular Chemistry and give a flavour of its exciting prospects for the future.

Suggested Further Reading

For an introduction to the concepts and principles that underpin Supramolecular Chemistry see: J.-M. Lehn, *Angewandte Chemie, International Edition in English*, **1988**, *27*, 89-112. For an overview of early Supramolecular Chemistry see: F. Vögtle, *Supramolecular Chemistry*, John Wiley and Sons, New York, 1991; for a personal view of the field see: J.-M. Lehn, *Supramolecular Chemistry: Concepts and Perspectives*, VCH, Weinheim, 1995. For more details on the macrocyclic effect see: E. Constable, *Macrocyclic Chemistry*, OCP, Oxford University Press.

2 Cation binding

In this chapter we will look at synthetic macrocyclic receptors for cations. Many of the receptors discussed bind cations *via* electrostatic ion-dipole interactions, and in some cases are enhanced by the formation of hydrogen bonds. A range of concepts will be introduced including the template effect, optimal spatial fit and preorganisation.

It should be noted that other non-macrocyclic ligands ranging from hexadentate ethylene-diaminetetracarboxylate ($EDTA^{4-}$) to metal coordinating proteins are also effective cation binding agents. They are, however, beyond the scope of this chapter.

2.1 Why bind cations?

Cations play many roles in *biological processes*. Concentration gradients of cations across cell membranes maintain potentials that are used to transport organic substrates into cells. They trigger muscle contraction and are involved in the transmission of nerve impulses. Metal cations are present at the active sites of many enzymes playing catalytic roles. Additionally, metal cations play vital roles in enzymes, stabilizing the polypeptide tertiary structure. They may also be transported in proteins (for example iron is transported in transferrin - see Fenton's Biocoordination Chemistry primer).

    Medically important metal complexes include complexes of paramagnetic lanthanide cations (e.g. Gd^{3+}) that are used as contrast agents in magnetic resonance imaging (MRI) of soft tissue (see chapter 7). Complexes of platinum (e.g. cis-platin) have been shown to coordinate to DNA, disrupting replication, and therefore hindering the growth of tumor cells (see Wilkins and Wilkins' Inorganic Chemistry in Biology primer).

    In *mining technology*, the design of receptors to selectively bind a particular metal ion in the presence of other cations is important in the extraction of a specific metal from a solution phase mixture (*e.g.* isolation of copper from copper ores containing mixtures of copper, iron, and cadmium).

    The development of sequestering agents for *toxic* or *pollutant* metals such as lead, cadmium, mercury or plutonium may be used in chelation therapy for poisoning victims or in systems designed to remove such pollutants from the environment. Such sequestering agents are used to remove excess iron from sufferers of haemophilia.

    There has therefore been a great deal of effort aimed at producing selective receptors for cationic guest species.

2.2 The synthesis of macrocycles

The majority of receptors in this Chapter (and indeed in this book) contain macrocyclic rings. We have already seen that Nature uses macrocyclic structures for metal ion binding in antibiotics such as valinomycin (Fig. 1.5). Additionally, *porphyrin* macrocycles are used to bind iron in haem groups, magnesium in chlorophyll a (**2.1**), a *corrin* macrocycle binds cobalt in vitamin B_{12} and a reduced form of a tetrapyrrolic macrocycle called a

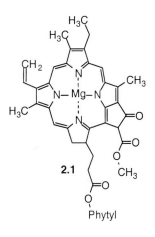

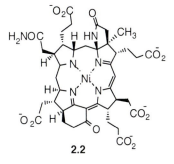

Fig. 2.1 Chlorophyll a (**2.1**) and factor F-430 (**2.2**).

corphin (Fig. 2.1) binds nickel in factor F-430 (**2.2**). We should therefore pause before we start looking at specific examples of the coordination properties of macrocyclic receptors and consider how macrocycles can be synthesized.

High dilution techniques

Macrocycle synthesis reactions carried out at *high dilution* favour cyclization over polymerization and can produce high yields of macrocyclic products. At high concentration polymerisation reactions dominate. The high dilution method is used in the synthesis of cryptands (the synthesis of [2.2.2]cryptand **2.4** is shown in Scheme 2.1). In a typical synthesis, solutions of each component are slowly added dropwise to a large volume of solvent, keeping the concentration of the intermediate low. Another technique involves attaching one reagent to a solid support (isolating it and preventing polymerization) and then exposing the support to the second reagent.

Coordination template effects

Metal ion templates are commonly used to increase yields of macrocycle formation. There are two classes of template effect - thermodynamic and kinetic (Scheme 2.2).

The *thermodynamic template effect* occurs when one product of a reaction is complementary to an added metal ion. In the absence of metal ion an equilibrium exists between the desired product and other species. However, on addition of the metal ion the desired product binds the metal ion, and the equilibrium is shifted in favour of the complex. This is illustrated in Scheme 2.2a) where a nickel-imine complex **2.5** is formed in > 70% yield, whereas in the absence of nickel the imine is not the major product of the reaction.

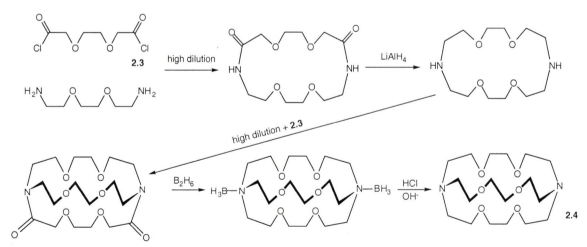

Scheme 2.1 Cryptands (Section 2.4) are synthesized using high dilution techniques.

The *kinetic template effect* is an organizational effect. The reactants wrap around the template, bringing their reactive sites into close proximity. Cyclization therefore becomes favoured over polymerization. This is shown in Scheme 2.2b) for the reaction of the nickel imine complex **2.5** formed in part a) with a bis-bromo compound forming macrocycle **2.6**. The kinetic template effect is used in the synthesis of phthalocyanins such as **2.7** (formed spontaneously on heating 1,2-dicyanobenzenes in the presence of metal ion templates, Scheme 2.3), crown ethers (Section 2.3), and sepulchrates (Section 2.7). In addition, metal cations have been used to direct the self-assembly of complex molecular architectures (see Chapter 6).

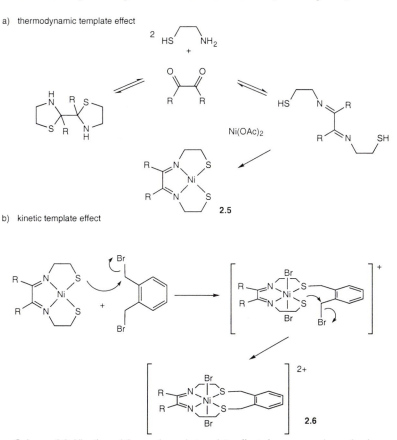

a) thermodynamic template effect

b) kinetic template effect

Scheme 2.2 Kinetic and thermodynamic template effects for macrocycle synthesis.

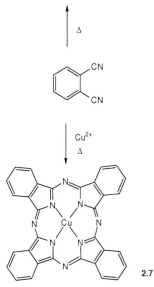

no macrocycle

Scheme 2.3 A metal ion template is required for phthalocyanin formation.

Pioneering work on the use of metal ions as templates for macrocycle synthesis was conducted by Curtis and co-workers in the 1960s. Reaction of $[Ni(en)_3]^{2+}$ with dry acetone gave a yellow crystalline product that was originally thought to be a simple Schiff base nickel (II) complex. However the product was found to be stable to very harsh conditions, such as boiling in alkali or acid, suggesting that it could not be the simple complex. It was eventually shown that the product was a macrocycle (**2.8**–Scheme 2.4).

Nelson and co-workers looked at Schiff base condensation reactions with diamines and diketones. They found that the condensation of 2,6-diacetylpyridine with 3,6-dioxaoctane-1,8-diamine was influenced by the nature of the of metal ion template (Scheme 2.5).

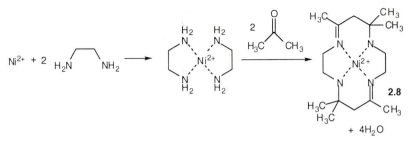

Scheme 2.4 Curtis synthesis of a Schiff base macrocycle

In the presence of smaller metal ions such as Mn^{2+}, Fe^{2+} or Mg^{2+} a $[1+1]$ condensation occurs (that is one molecule of diamine condenses with one molecule of diketone) forming the small macrocycle **2.9**. However, in the presence of larger cations (Ba^{2+} or Pb^{2+}) a $[2+2]$ condensation occurs producing a larger macrocycle **2.10** (this reaction proceeds through an intermediate containing two metal ions). This clearly illustrates that the choice of metal ion template is crucial in macrocycle synthesis.

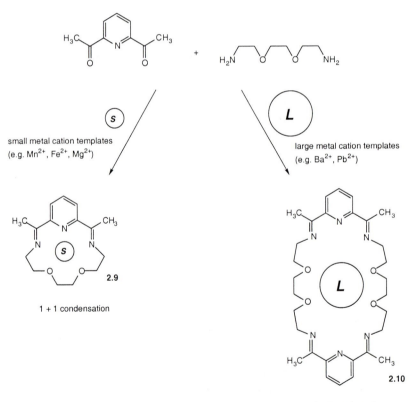

Scheme 2.5 The size of the macrocyclic product is influence by the size of the metal cation template used.

De-metallation

One disadvantage of template synthesis over high dilution methods is that once the macrocycle has been synthesized the metal ion template must be removed (without destroying the macrocyclic framework). In some cases this is virtually impossible (see sephulcrates in Section 2.7). However there are several techniques commonly used to de-metallate macrocycles:

(a) For weakly coordinating metals it is sometimes possible to dissolve the macrocycle–metal complex in an organic solvent and then wash the organic phase with water. The water will effectively solvate the metal ion and remove it.

(b) Addition of acid to macrocycles containing chelated amine groups can cause de-metallation. The acid protonates the amine groups so 'using up' their electron pair in an N-H bond. This electron pair is then unavailable for metal coordination and the macrocycle will de-metallate. Removal of metals from Schiff base macrocycles may involve reduction of the Schiff base to its corresponding amine followed by protonation.

(c) Addition of a strongly coordinating competitive ligand to the macrocycle-metal complex in solution can sequester the metal ion from the macrocycle. For example, addition of CN^- to Sauvage's catenates and metallated trefoil knots (see Section 6.3) removes the coordinated copper(I) metal ions. Other strongly coordinating ligands such as $EDTA^{4-}$ have also been used in this procedure.

(d) By changing the oxidation state of a metal it is sometimes possible to switch from a kinetically inert complex to a labile one. For example inert complexes containing Cr(III) or Co(III) may be reduced to their Cr(II) or Co(II) oxidation states that are more labile. One of the other de-metallation procedures described above may then be used.

The following sections of this chapter describe specific examples of cation coordinating macrocycles beginning with the discovery of the crown ethers by Charles Pedersen.

2.3 Crown ethers

The genesis of supramolecular chemistry may be traced back to the pioneering work of Charles J. Pedersen. Pedersen was an industrial chemist working for DuPont in New Jersey, USA. He was interested in making multidentate ligands for copper and vanadium and in 1960 was attempting to synthesize bis[2-(o-hydroxy-phenoxy)ethyl]ether **2.11** (Scheme 2.6) when he made a fascinating chance discovery.

Whilst attempting to purify the expected product he isolated some white crystals in approximately 0.4% yield. He found that addition of sodium salts increased the solubility of these crystals in methanol by a very large factor. The crystals were characterized by elemental analysis and mass spectrometry and found to be a macrocycle (**2.12**) (Fig. 2.2) formed due to the presence of catechol impurity and a sodium template ion in the second step of the synthesis. The systematic names to these compounds are rather long, and this prompted Pedersen to give this class of compound the trivial name '*crown ether*' (because of their 'crown-like' conformation in the solid phase) and this particular compound the name dibenzo[18]crown-6. [18] refers to the number of atoms in the macrocycle and 6 to the number of oxygen atoms in the ring.

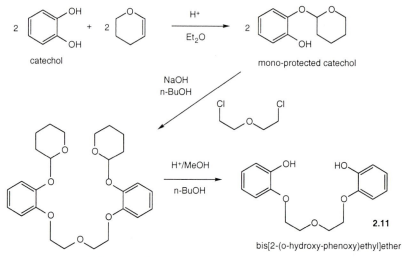

Scheme 2.6 Intended synthesis of Pedersen's ligands

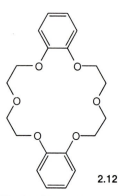

Fig. 2.2 Dibenzo[18]crown-6 (or 2,3,11,12-dibenzo-1,4,7,10,13,16-hexaoxacyclooctadeca-2,11-diene).

Pedersen observed from a space filling model that a sodium ion can sit in the cavity of the crown, held by attractive electrostatic ion-dipole interactions between the alkali metal cation and the six oxygen donor atoms in the polyether rings. He recognized that the increased solubility of the macrocycle in hydroxylic solvents in the presence of sodium cations was due to the crown binding a sodium cation. This molecule was the first synthetically produced receptor for Group 1 metal cations. Pedersen was awarded the Nobel prize for this work in 1987 (jointly with Profs Jean-Marie Lehn and Donald J. Cram, some of whose work is discussed later in this Chapter).

[18]crown-6 **2.13**

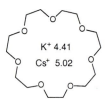

[21]crown-7 **2.14**

Fig. 2.3 Stability constants (log K) for sodium, caesium and potassium in methanol. Optimal spacial fit: [18]crown-6 selectively binds potassium cations whereas [21]crown-7 is selective for caesium cations.

Table 2.1 Cation diameters for the alkali metals and crown ether cavity sizes

Cation	Diameter (Å)	Crown Ether	Cavity diameter (Å)
Li^+	1.36	[12]crown-4	1.2–1.5
Na^+	1.94	[15]crown-5	1.7–2.2
K^+	2.66	[18]crown-6	2.6–3.2
Rb^+	2.94		
Cs^+	3.34	[21]crown-7	3.4–4.3

A relationship exists between the cavity size, cationic radius and stability of the resulting complex (Table 2.1). The better the fit of the cation into the crown, the stronger the complex formed. This concept is referred to as *optimal spatial fit*. Frensdorff determined the stability constants of various crown ethers with cations in methanol by potentiometry and found, for example, that [18]crown-6 (**2.13**) forms the most stable complexes with potassium cations whereas [21]crown-7 (**2.14**), a larger crown, binds caesium more strongly than potassium (Fig. 2.3).

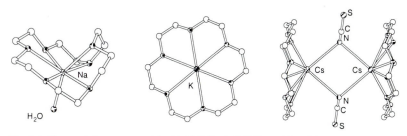

Fig. 2.4 The crystal structures of the NaCNS.H$_2$O, KCNS, and CsNCS complexes of [18]crown-6 (**2.13**).

There is also structural evidence for this concept of optimal spatial fit. The crystal structures of the potassium, sodium and caesium complexes of [18]crown-6 are illustrated (Fig. 2.4). The cavity size of [18]crown-6 is complementary for potassium ions (Table 2.1), and therefore [18]crown-6 forms a 1:1 complex with a potassium cation sitting perfectly in the middle of the macrocycle. Complexes formed between [18]crown-6 and other alkali metal cations, however, are considerably less stable than the potassium complex (Fig. 2.3). For smaller cations, such as sodium, the crown ether distorts, wrapping itself around the metal in an attempt to maximize the electrostatic interactions, but at the same time increasing the strain of the ligand. Larger cations such as caesium have to perch above one face of the macrocycle because they are too large to fit into the cavity. This mode of binding decreases the electrostatic interactions between ion and crown.

Crown ethers may also form complexes of different stoichiometry with alkali metal cations. The crystal structure of K$^+$–(benzo[15]crown-5)$_2$ (Fig. 1.14) revealed that the potassium cation is sandwiched between two crown ether molecules (i.e. it is too large to fit in the cavity of the macrocycle). The inverse situation also occurs, with large crowns being capable of binding more than one cation at once. Dibenzo[24]crown-8 (**2.15**), for example, can encapsulate two sodium cations simultaneously (Fig. 2.5).

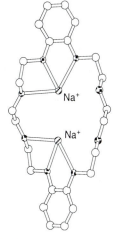

Fig. 2.5 The crystal structure of 2Na$^+$–[24]crown-8 (**2.15**).

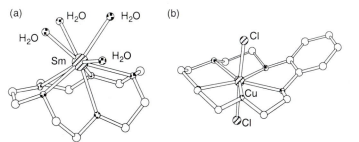

Fig. 2.6 Crown ethers also form complexes with other metals: a) a samarium-[15]crown-5 complex **2.16** and b) a copper-[15]crown-5 complex **2.17**.

Crown ethers may also form complexes with transition metals and lanthanide cations. A samarium complex of [15]crown-5 (**2.16**) is shown in Fig.2.6a. A copper complex of benzo[15]crown-5 (**2.17**) is shown in Fig. 2.6b. The copper adopts a 7-coordinate geometry with the crown ether coordinating equatorially and two chloride ligands axially.

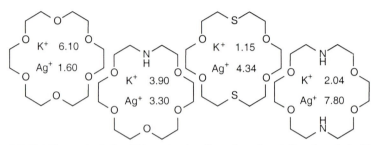

Fig. 2.7 Stability constants (log K) for potassium (in methanol) and silver (in water) with oxa- , oxaaza- and oxathia- crown ethers.

The affinity of crown ethers for transition metals can be enhanced by replacing oxygen atoms with softer donor atoms such as sulfur or nitrogen. Figure 2.7 shows the binding constants of a variety of crown ethers with potassium and silver. The binding constant of the silver cation increases at the expense of the affinity of the macrocycle for harder cations such as potassium. This effect can be explained by the *'hard-soft acid-base' principle* with which the reader should be familiar.

Crown ethers do not, however, only bind simple metal ions, they can also bind ammonium and alkylammonium cations. [18]Crown-6 is a complementary receptor for primary ammonium cations as it possesses three oxygen atoms correctly oriented to form three *hydrogen bonds* with the guest (Fig. 2.8). In this case, therefore, the ammonium ion is being held by a combination of electrostatic interactions and hydrogen bonds. [15]Crown-5, however, does not possess C_3 symmetry and is therefore not complementary to the guest (the oxygen atoms are not in the correct orientation to form three strong hydrogen bonds). This *mismatch in symmetry* is reflected in much weaker complexes being formed between the smaller crown ether and ammonium cations.

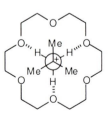

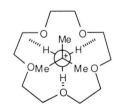

ammonium guest

strong complex
ideal arrangement of hydrogen bond donors and acceptors

weaker complex
mismatch in geometry of hydrogen bond donors and acceptors

Fig. 2.8 [18]Crown-6 has C_3 symmetry and is ideally suited for binding ammonium and primary alkyl ammonium cations. Ammonium complexes of [15]crown-5 are less stable due to the mismatch in symmetry.

These principles can be extended to other crown ethers. For example, the guanidinium cation (see section 3.6) has been used as a template in the synthesis of benzo[27]crown-9 (**2.18**). The cation forms six hydrogen bonds with the poly-glycol chain causing the reactive ends to come into close proximity (Fig. 2.9).

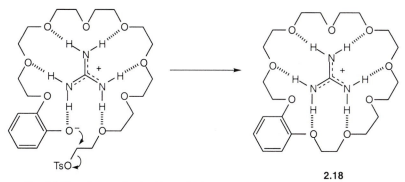

2.18

Fig. 2.9 Template synthesis of benzo[27]crown-9 (**2.18**) using guanidinium cations.

Crown ethers can also be functionalized with pendant arms containing additional coordinating groups and are then known as *lariat* crown ethers. Lariat crowns were designed as carrier species for cations across lipophilic membranes, because they possess higher binding constants than regular crowns, but remain kinetically labile (i.e. the rates of complexation and decomplexation are fast – a requirement for efficient transport) due to the flexibility of the arm. The cation is bound by the oxygen atoms in the crown and also by the pendant arm (by groups such as OMe). Scheme 2.7 shows the two possible ways of attaching a pendant arm to the crown ether skeleton, a) *via* a carbon atom (**2.19**) and b) using an aza-crown ether with a pendant arm attached to the nitrogen atom (**2.20**). An example of a carbon pivot lariat crown (**2.21**) that is selective for lithium (on the basis of size) is shown in Fig. 2.10.

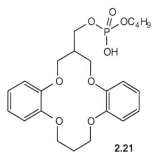

2.21

Fig. 2.10 A Li$^+$ selective lariat crown ether (**2.21**).

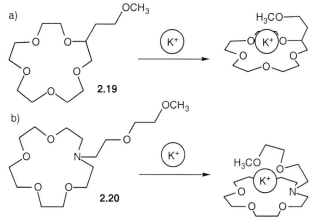

Scheme 2.7 (a) Carbon and (b) nitrogen pivot lariat crown ethers binding potassium cations.

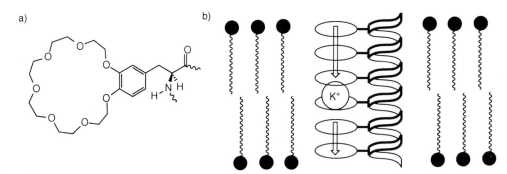

Fig. 2.11 (a) [21]crown-7 L-phenylalanine is incorporated into an α-helical peptide chain (b) when the peptide is incorporated into a membrane bilayer it allows potassium ions to pass through.

Crown ethers are now being used by supramolecular chemists in increasingly sophisticated ways. For example, Voyer recently produced synthetic cation channels mimicking those used in biology to span cell membranes, by attaching crown ethers to a polypeptide backbone. A 21 amino acid peptide consisting of 15 hydrophobic L-leucine residues and six [21]crown-7-L-phenylalanines was synthesized. The crown ethers were placed in the chain such that when the peptide coils up in an α-helical conformation the crown ethers align to form a channel (Fig. 2.11). The channel was incorporated into synthetic lipid bilayers and shown by pH techniques to allow ions to pass through. [21]crown-7 was chosen because it is too large to form strong complexes with alkali metal cations. If a smaller crown ether, such as [18]crown-6 were used, a strong complex would be formed with potassium ions which would block the channel.

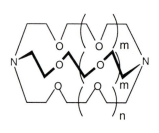

m=0, n=1 [2.1.1] **2.22**
m=1, n=0 [2.2.1] **2.23**
m=1, n=1 [2.2.2] **2.4**

Fig. 2.12 Cryptands [2.1.1], [2.2.1]) and [2.2.2].

2.4 Cryptands

In 1969 Lehn and co-workers first reported the synthesis of a new class of molecule which they named '*cryptands*'. These receptors effectively moved crown ethers into the third dimension. Cryptands are cage like bicyclic molecules, synthesized by high dilution techniques (Section 2.2), and are named according to the number of oxygen atoms in each nitrogen-nitrogen linker (Fig. 2.12). They have been found to complex group 1 and 2 metal cations with very large stability constants, higher than those of analogous crown ethers. In *methanol* [2.2.2]cryptand **2.4** binds potassium cations with a stability constant (log K) of 10.4 (over four orders of magnitude higher than [18]crown-6, logK=6.10). The complex of a cryptand is referred to as a cryptate. The stability constants for [2.1.1] (**2.22**), [2.2.1] (**2.23**), and [2.2.2] (**2.4**) cryptands in *water* are shown in Fig. 2.13. The highest stabilities are observed for Li$^+$[2.1.1], Na$^+$[2.2.1] and K$^+$[2.2.2] (Fig 2.14) cryptates which correspond to complementary cryptand cavity-cation size.

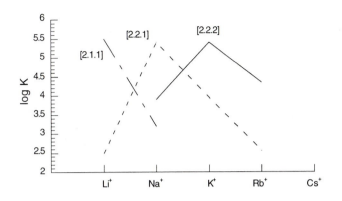

Fig. 2.13 Stability constants for [2.1.1], [2.2.1], and [2.2.2] cryptands with alkali metal cations in water (for points not plotted log K<2).

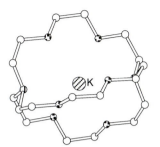

Fig. 2.14. The crystal structure of the potassium iodide complex of [2.2.2]-cryptand.

The high stability constants have been attibuted to a number of factors. Once bound in the cryptand cavity the cation is poorly solvated leading to a positive entropy effect, as solvent is released on binding. There is also less reorganization of the receptor's structure required to coordinate the metal ion than for the crown ethers, thus giving a favourable enthalpic contribution.

2.5 Spherands: preorganized receptors

The crystal structures of uncomplexed [18]crown-6 and [2.2.2]cryptand show that their binding sites are not completely convergently arranged (Fig. 2.15). Complexation in solution must therefore involve some degree of ligand rearrangement. Donald Cram recognized that ligands which were rigid and contained binding sites fixed in an octahedral arrangement around an enforced cavity should show enhanced binding over flexible ligands. With the help of CPK molecular models he designed a ligand containing an enforced spherical cavity (**2.26**: Fig. 2.16). He gave this family of preorganized ligands the name *spherand*, and the complexes were termed spheraplexes.

a)

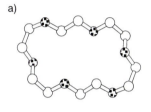

b)

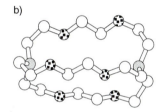

Fig. 2.15. Crystal structures of a) uncomplexed 18-crown-6 and b) [2.2.2] cryptand.

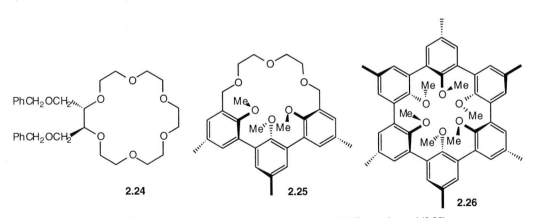

2.24   **2.25**   **2.26**

Fig. 2.16 Increasing preorganization from crown ether (**2.24**) to spherand (**2.26**).

Table 2.2 Spherand stability constants (M^{-1}) at 25°C in $CDCl_3$ saturated with D_2O

Host	Li$^+$	Na$^+$
2.24	4.1×10^4	1.4×10^6
2.25	1.9×10^5	8.7×10^8
2.26	$> 7 \times 10^{16}$	1.2×10^{14}

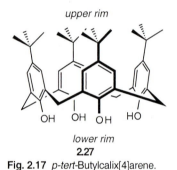

upper rim

lower rim
2.27
Fig. 2.17 *p-tert*-Butylcalix[4]arene.

Consider receptors **2.24-2.26**. As the receptor becomes more rigid, the binding strength increases dramatically. In addition, the rigid spherand host (**2.26**) is completely unable to complex cations larger than sodium, unlike crown ether **2.24** and 'half-spherand' **2.25**. Cram formalized these observations in the principle of *preorganization* which states that "the more highly hosts and guests are organized for binding and low solvation prior to their complexation, the more stable will be their complexes." Preorganisation is therefore a central determinant of binding power. As for crown ethers, the spherand cavity size can be tuned to yield selectivity for different cations. For example spherand **2.26** displays a remarkable selectivity for Na$^+$ and Li$^+$ over other Group I metal cations. K$^+$ is not bound at all as it is too large to fit in the spherand cavity. Consequently spherands can be used to remove Li$^+$ and Na$^+$ impurities to obtain ultra pure samples of potassium salts. As expected these systems show slow complexation–decomplexation kinetics.

2.6 Calixarenes as cation binding agents

Calix[*n*]arenes are a family of synthetic macrocyclic receptors consisting of cyclic arrays of *n* phenol moieties linked by methylene groups. They are formed from the base catalysed condensation of *p-tert*-butylphenol and formaldehyde. The chemistry of calix[*n*]arene formation has been studied in great detail by C. David Gutsche and co-workers. They found that by altering the reaction conditions, differently sized calix[*n*]arenes could be produced.

Calix[4]arenes (Fig. 2.17) may adopt one of four conformations: cone, partial cone, 1,2-alternate or 1,3-alternate (Fig. 2.18). *p-tert*-Butylcalix[4]arene **2.27** exists in a cone conformation that is stabilized by an array of hydrogen bonds between the phenolic OH groups at the lower rim of the macrocycle. Deprotonated calixarenes are capable of binding alkali metal cations at the lower rim, and can even act as carriers, transporting these guests through model membrane systems.

The cone conformation is particularly useful because its 'bucket' shape makes an attractive three dimensional template from which to hang ligating groups. Groups may be appended to the lower rim or, after removal of the *tert*-butyl groups, to the upper rim of the calixarene. For example, ester groups may be attached to the lower rim of **2.27** producing the tetraester compound **2.28** in the cone conformation. UV spectroscopic techniques show that this receptor is selective for sodium cations over other putative Group 1 guest species (Fig. 2.19). The sodium is coordinated to eight oxygen atoms at the lower rim of the calixarene (Fig. 2.20). In contrast to crown ethers, the coordinating oxygen atoms are not integral parts of a macrocyclic structure but rather, are appended to it.

The aromatic cavities of calixarenes are also capable of coodinating to guest species. An example of a cation included in a calixarene cavity is shown in Fig. 2.21a, in which a caesium cation is held in the cavity of *p-tert*-butylcalix[4]arene **2.26** *via* π-*cation interactions* (a molecule of acetonitrile is also coordinated to the metal cation). These π-cation interactions are believed to arise from favourable electrostatic interactions between the electron deficient cation and the electron rich aromatic ring.

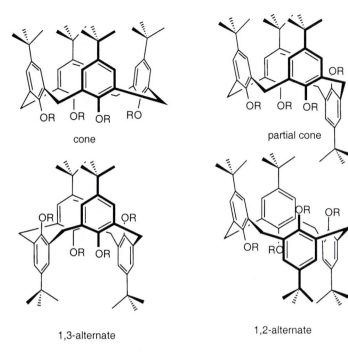

cone

partial cone

1,3-alternate

1,2-alternate

Fig. 2.18 Conformations adopted by calix[4]arenes.

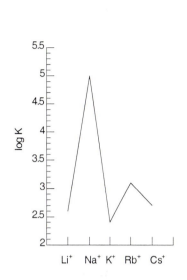

Fig. 2.19 Log K values for Group 1 metal cation complexes of *p-tert*-butylcalix[4]arene tetra ester **2.28** in methanol.

This π-cation interaction has also been used to produce a model for the selection filter present in *potassium channel proteins*. This filter is believed to consist of a square planar array of four converging aromatic amino acid (tyrosine) residues, a structure not dissimilar to calix[4]arene. Calix[4]tube **2.29** (Fig. 2.21b) is a bis-calix[4]arene linked lower-rim to lower rim *via* four ethylene bridges. Potassium cations pass through the calixarene cavity and are bound by the eight phenolic oxygens in the middle of the tube (Fig. 2.22). Uptake of other Group 1 metal cations by the tube is insignificant.

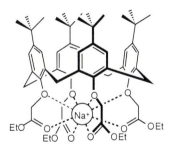

Fig. 2.20 The sodium complex of **2.28**.

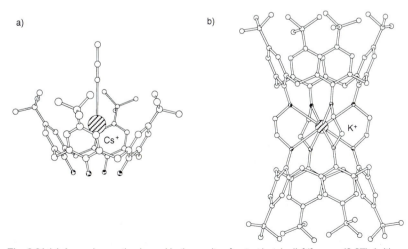

Fig. 2.21 (a) A caesium cation bound in the cavity of *p-tert*-butylcalix[4]arene (**2.27**) (with an acetonitrile molecule) and (b) the structure of calix[4]tube (**2.29**) - potassium complex.

2.7 Sepulchrates

Cryptand-like molecules may also be synthesized using metal ions as templates. The first example of this type of synthesis was reported in 1968 by Boston and Rose. The tris(dimethylglyoximato)cobalt(III) complex **2.30** reacts with two equivalents of BF_3 forming tripodal FBO_3 bridgeheads and encapsulating the cobalt cation **2.31** (Scheme 2.8).

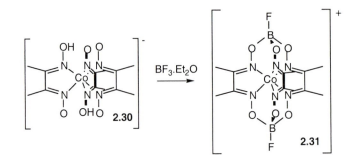

Scheme 2.8 Templated cryptand synthesis.

Sargeson and co-workers extended this synthetic strategy to the synthesis of octaazacryptands e.g. **2.32** (Scheme 2.9). Sargeson gave these complexes the name *sepulchrate* because of the extremely high stability of the cobalt complex formed. Analogous compounds containing carbon bridgeheads are known as *sarcophagines* e.g. **2.33** (Fig. 2.23).

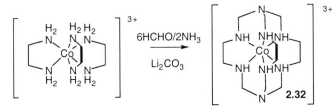

Scheme 2.9 Synthesis of sepulchrate **2.32**.

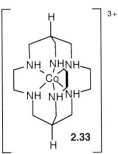

Fig 2.23 The cobalt(III) complex of sarcophagine **2.33**.

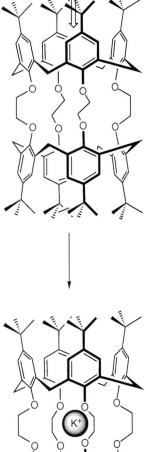

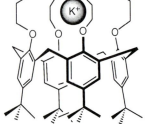

Fig. 2.22 Potassium encapsulation in calix[4]tube **2.29**.

2.8 Siderophores

Iron is an essential element for life (see David Fenton's Biocoordination Chemistry primer). It is, however, particularly difficult for fungi and bacteria to obtain because the hydrolysis of iron(III) limits its concentration to 10^{-18} moldm^{-3} at pH7. In order to scavenge the available iron, fungi and bacteria therefore produce complexing agents, called *siderophores* (from the Greek: σιδεροσ (iron) and; φορενζ (carrier)) which they release and then reabsorb as an iron complex *via* an active transport mechanism. The siderophore forms a six-coordinate complex with the iron(III) *via* hydroxamate or catecholate oxygen atoms (Fig. 2.24).

hydroxamate

catecholate

Fig. 2.24 Hydroxamate or catecholate groups chelate iron(III) in most siderophores.

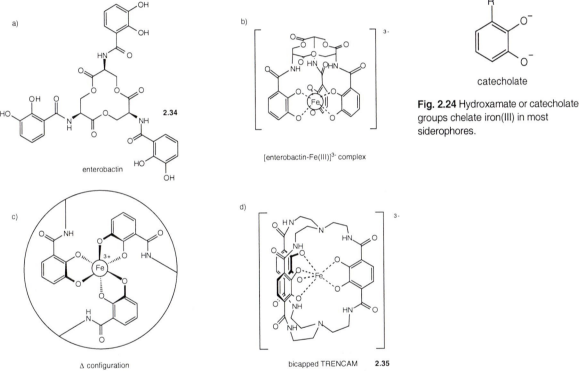

Fig. 2.25 (a) The structure of enterobactin **2.34**, (b) enterobactin-iron(III) complex , (c) the coordination environment of the bound iron, (d) bicapped TRENCAM **2.35**, a synthetic model for enterobactin.

Enterobactin **2.34** is a bacterial siderophore that forms extremely stable complexes with iron(III) (Fig. 2.25), in fact the stability constant of the [enterobactin-Fe]$^{3-}$ complex in aqueous solution is 10^{52}moldm^{-3}. It consists of a cyclic triester linked to three catechol groups *via* amide bonds. The triester scaffold imposes a Δ configuration of catecholates around the metal. Once inside the bacterium an enzyme breaks the enterobactin skeleton, releasing the iron.

Kenneth Raymond and co-workers at the University of California, Berkeley have produced a number of siderophore analogues based on catechol. One such cryptand-like analogue is shown in Figure 2.25d). The catechol cage molecule

2.35 was synthesized using both high dilution techniques (to form the free ligand) and an iron templated approach (to produce the iron-cage complex). These types of artificial siderophores have carefully tailored recognition sites, and consequently show extremely strong binding with high selectivity.

The coordination chemistry of Fe(III) is similar to that of Pu(IV). Thus artificial siderophores may find applications as sequestering agents for Pu(IV) leading to treatments for plutonium contamination of both people and the environment.

Summary and conculsions

The aim of this chapter has been to demonstrate how receptor structure, complementarity with the substrate, and preorganization influence the strength and selectivity of complexes formed between polydentate receptors and cations. There are many examples of synthetic receptors for cations and this chapter does not attempt to provide a comprehensive survey of the area. Rather it gives a taste of a wide range of systems which will hopefully encourage the reader to consult the literature cited in the section below.

Suggested further reading

For an interesting account of the discovery of crown ethers: C.J. Pedersen, *Angewandte Chemie, International Edition in English,* **1988**, 27, 1021-1027 and for an overview of the synthesis and coordination properties of macrocycles (including crown ethers and cryptands) see: B. Dietrich, P. Viout, and J.-M. Lehn, *Macrocyclic Chemistry,* VCH, Weinheim, 1993. For a practical approach to macrocycle synthesis see: *Macrocycle Synthesis*, Ed. D. Parker, Oxford University Press, Oxford, 1996. For an account of the chemistry of spherands and cavitands see: D.J. Cram and J.M. Cram, *Container Molecules and Their Guests, Monographs in Supramolecular Chemistry,* Ed. J.F. Stoddart, The Royal Society of Chemistry, Cambridge, 1994 and for further details about calixarenes see C.D. Gutsche, *Calixarenes, Monographs in Supramolecular Chemistry,* Ed. J.F. Stoddart, The Royal Society of Chemistry, Cambridge, 1989. For a review of siderophores and their model compounds see: K.N. Raymond, *Coordination Chemistry Reviews,* **1990**, 105, 135-153.

3 Anion binding

In comparison to cation coordination chemistry, the idea of constructing hosts to bind specific negatively charged guests is a recent development. The importance of anions, however, is increasingly recognized, and designing receptors for their recognition is a vigorously active research area.

3.1 Properties of anions: receptor design principles

Anions have a variety of special features that must be addressed with effective receptor design if they are to be strongly bound.

Charge: The defining feature of an anion is its negative charge. Electrostatic interactions can therefore play an important role in strengthening anion coordination and many receptors utilise them.

Size: Anions are larger than their isoelectronic cations (Fig. 3.1) and therefore recognition cavities must be large enough for guest encapsulation.

pH dependence: Unlike simple metal cations, many anions (e.g. carboxylates, phosphates, sulphates) only exist over a limited pH range. At low pH they become protonated and consequently lose their negative charge. This is especially important if anion recognition is to occur in water.

Solvation: The solvation of anions depends on the three factors above, and plays an important role in controlling binding selectivity. The degree of aqueous solvation of an anion is reflected by the *Hofmeister series*, which indicates increasing hydrophobicity (i.e. low aqueous solvation: Fig. 3.2). In a hydrophobic binding site, which solvent cannot enter, the less hydrated anions are bound more strongly. In a cavity which is accessible to polar solvents, however, the more hydrated anions are often more strongly held.

Geometry: Anions exhibit a range of geometries (Fig. 3.3), which challenge the molecular designer to create a complementary binding site.

Examples of the application of these design principles will be illustrated throughout the course of this Chapter.

3.2 Why bind anions?

Anions play essential roles in many processes, both chemical and biological, and this makes their strong, selective recognition an area of intense interest.

Chemically, anions play various roles, acting for example as catalysts and bases. The use of a receptor to bind an anion can alter its reactivity. Anion receptors may also assist in the separation of complex chemical mixtures.

Environmentally, anions can pose severe pollution problems. Nitrate anions from fertilisers run off agricultural land into the water supply where they have a fertilising effect, leading to eutrophication (excessive plant/algal growth) that disrupts aquatic life cycles. Pertechnetate anions are a toxic,

Na^+	0.95Å	F^-	1.36Å
K^+	1.33Å	Cl^-	1.81Å
Rb^+	1.48Å	Br^-	1.95Å
Cs^+	1.69Å	I^-	2.16Å

Fig. 3.1 Ionic radii of isoelectronic pairs of cations and anions.

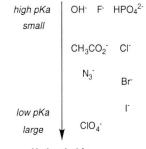

Fig. 3.2 The Hofmeister Series.

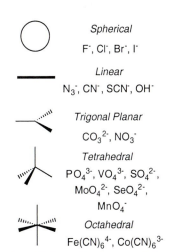

Fig. 3.3 Examples of anions which possess different geometries.

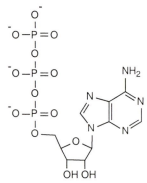

Fig. 3.4 Adenosine triphosphate (ATP) is a biologically important anion.

radioactive by-product of the nuclear power industry. Selective binding, extraction and sensing of such anions are, therefore, important goals.

Biologically, 70% of all enzyme substrates are negatively charged. Adenosine triphosphate (ATP), the free energy source of life, is a tetraanion at physiological pH (Fig. 3.4). Deoxyribonucleic acid (DNA) is also a polyanion, containing phosphate esters along its ribose backbone. Anion recognition in biology is remarkably subtle, for example all cells must distinguish between anions as similar as SeO_4^{2-}, SO_4^{2-} and MoO_4^{2-}. Anionic phosphate nucleotides labelled with radioactive ^{32}P have long been used for tracking reaction progress in biochemical investigations.

Medically, anions play key roles in many diseases. Cystic fibrosis, a genetically inherited disease, is caused by misregulation of chloride channels, yet current methods of chloride analysis are unsuitable for medical use.

There are therefore many compelling reasons for investigating anion coordination. The applications of supramolecular chemistry are discussed in more detail in Chapter Seven. This Chapter, however, will focus on the use of receptor design principles to achieve selective anion recognition.

3.3 Recognition using electrostatic interactions

Perhaps the simplest and most obvious way of binding an anion is to use a positive charge to provide a strong *electrostatic ion–ion interaction*. By extension, an *array* of positively charged groups should be even more effective. Building a number of positively charged groups into a receptor, however, causes a design problem, as the positive charges will tend to repel one another. The charges are therefore often constrained by a rigid or cyclic system. This has the added advantage of creating a binding cavity.

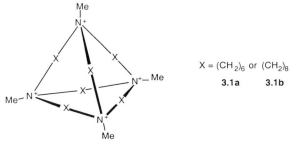

Fig. 3.5 Receptor with an electron deficient cavity suitable for anion binding.

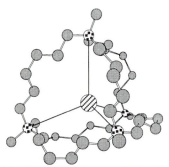

Fig. 3.6 Crystal structure of **3.1a** with bound iodide anion.

Schmidtchen and co–workers reported a macrotricyclic receptor for anionic guests with positively charged, quaternized nitrogen atoms at its corners (Fig. 3.5). Receptor **3.1a** was prepared as its tetraiodide salt and crystallography located one of the iodide ions as being bound at the centre of the cavity, equidistant from the four positively charged nitrogen atoms (Fig. 3.6). The receptors also formed 1:1 complexes with anionic guests in aqueous solution, the strengths of which are shown in Table 3.1. Receptor **3.1a** binds anions more strongly than receptor **3.1b** as its smaller diameter (4.6Å rather than 7.6Å) leads to a better fit of the anions into the cavity. Bromide is bound more strongly than chloride or iodide by this smaller host because it has the best steric fit. The larger host, however, forms the strongest complex with the larger iodide anion. This shows how the structure of the receptor directs *size selectivity*. In addition, **3.1a** fails to complex large anions, such as *p*–nitrophenolate, which the larger **3.1b** can bind.

There is, however, a problem with positively charged anion receptors: they possess a counter anion, which may act as a competitor for the binding site. Schmidtchen and his co–workers addressed this problem by synthesising zwitterionic hosts such as **3.2** (Fig. 3.7). This receptor has no net charge, and consequently there is no competitive counter anion. Nmr titrations (Section 1.4) showed that as a consequence, receptor **3.2** bound halide anions more strongly in aqueous solution than **3.1a** (Table 3.1).

Dipolar bonds can also be used for electrostatic anion recognition. They must, however, be well aligned and properly oriented for recognition to occur. Receptor **3.3** has its dipoles convergently focussed into a halide binding recognition site below the macrocyclic ring (Scheme 3.1). Binding constants were measured in a low polarity solvent mixture (CHCl₃/MeOH, 98/2), which does not disrupt electrostatic forces. In spite of this, the binding constants were still relatively small (K= 65 M⁻¹ for Cl⁻ binding). The recognition is, therefore, much weaker than for positively charged receptors **3.1** and **3.2**. Receptor **3.3** can, however, also bind cations at the same time as anions, through interactions with its electronegative oxygen atoms. The topic of ion pair binding will be discussed in more detail in Chapter Four.

Table 3.1 Binding constants (K[M⁻¹])for receptors **3.1** and **3.2** with halide anions in H₂O.

	3.1a	3.1b	3.2
Cl⁻	50	-	270
Br⁻	1020	100	2150
I⁻	500	290	6480

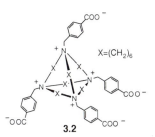

Fig. 3.7 A net neutral receptor eliminates counter–anion competition.

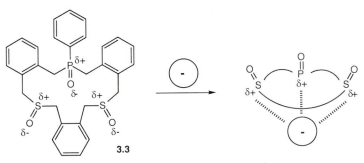

3.3

Scheme 3.1 Dipolar receptor for halide anions.

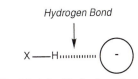

Fig. 3.8 Anion binding through hydrogen bond formation.

3.4 Recognition using hydrogen bonds

Simple electrostatic interactions, however, are not the only way of binding anionic guests. Any anion can be considered as an electron pair donor and it will therefore interact with a suitable electron pair acceptor. Perhaps the simplest such electron pair acceptor is an electropositive hydrogen atom capable of forming a hydrogen bond (O–H, N–H etc.) (Fig. 3.8). Consequently, *arrays of hydrogen bonding groups* can be used to bind anions.

As with dipoles, it is important that hydrogen bonds are correctly oriented, with the hydrogen atoms arranged in a *convergent* manner. Receptor **3.4** was the first anion host to function only through hydrogen bonds, with three convergent amide protons pointing into a cavity (Fig. 3.9). Nmr provided evidence that this receptor bound fluoride anions in dimethylsulphoxide (DMSO) solution.

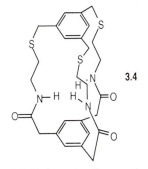

Fig. 3.9 Hydrogen bonding receptor for fluoride anions.

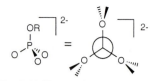

Fig. 3.10 The tetrahedral symmetry of phosphates makes them ideal guests for receptors with trigonal geometry

The directionality of hydrogen bonds introduces the possibility of designing receptors with *specific shapes*, which are capable of differentiating between anionic guests with different geometries or hydrogen bonding requirements. Trigonal receptors, for example, are of particular interest due to their complementarity with organic phosphate anions, which have a trigonal array of hydrogen bond acceptor groups (Fig. 3.10). Phosphate anions have great biological importance. Receptor **3.5** (Fig. 3.11) binds phosphonate and phosphate anions strongly, even in competitive hydrogen bonding solvents such as methanol, which can itself act as both a hydrogen bond donor and acceptor. This illustrates the principle of *complementary steric fit* within a recognition site.

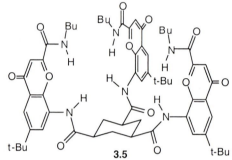

Fig. 3.11 Trigonal receptors can bind phosphate anions strongly in competitive solvents.

Urea groups are particularly effective neutral hydrogen bonding motifs for anion coordination. The hydrogen atoms are oriented in a *bidentate* manner, perfect for the formation of hydrogen bonds with complementary carboxylate or phosphate anions. Receptor **3.6** binds bidentate anions in dimethylsulphoxide (DMSO) solution (Fig. 3.12), with the strength of the complex depending on the basicity of the guest (Table 3.2).

Table 3.2 Bidentate guests for receptor **3.6**.

Guests	pK_b	K (M^{-1}) [DMSO]
OPO$_3$H$^-$	13	30
PO$_3$H$^-$	12	140
CO$_2^-$	10	150
PO$_3^{2-}$	7	2500

The lower the pK_b of the guest, the higher its basicity, and the more strongly it will accept a proton. Host **3.6** functions through 'donation' of protons *via* hydrogen bonds, and therefore forms stronger interactions with the more basic anions, binding them more effectively.

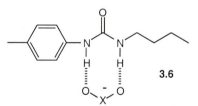

Fig. 3.12 Receptor **3.6** is complementary for bidentate guests.

Another approach to anion binding through hydrogen bonding does not use amide protons at all, but instead uses the pyrrolic protons of a calix[4]pyrrole macrocycle (**3.7**) (Fig. 3.13). The crystal structure of the free macrocycle has a 1,3-alternate conformation with alternating pyrrole rings oriented 'up' and 'down' (Fig. 3.14). On binding a halide anion, however, crystallography and nmr experiments indicated that the pyrrole rings align with all four N–H groups convergent, forming hydrogen bonds with the bound guest (Fig. 3.15).

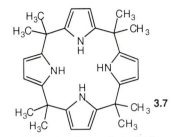

Fig. 3.13 *meso*–Octamethylcalix[4]pyrrole.

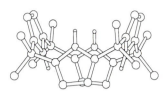

Fig. 3.14 Crystal structure of free *meso*–octamethylcalix[4]pyrrole (**3.7**).

3.5 Recognition using Lewis acidic hosts

As discussed above, anions are capable of donating an electron pair to some extent, allowing the use of hydrogen bonds to bind these guests. *Lewis acids*, however, can also accept an electron pair, and this has led to the development of a rich variety of anion receptors based on heteroelement chemistry.

Organoboron compounds, for example, only contain six electrons in the boron valence shell, rather than the full complement of eight. They are consequently ideal Lewis acids, capable of accepting electron pairs from anionic guests into their vacant orbitals. The first host of this type was the simple ligand, **3.8** reported by Shriver and Biallis in 1967. As such, it was one of the first designed receptors for anions. This *chelating* ligand can be thought of as an equivalent to ethylene diamine, only instead of donating two electron pairs to a metal ion, it is capable of accepting two electron pairs from an anion (Fig. 3.16). The binding of methoxide anions was investigated, and when **3.8** was compared with non–chelating, mononuclear boron trifluoride (BF_3), a *chelate effect* was observed. Subsequently, Katz synthesized receptor **3.9** which has two boron atoms rigidly held on an aromatic framework, leading to a degree of *preorganisation* (Fig. 3.17). In particular, 'hard' anions such as hydride and fluoride were bound, as they are electron rich and strong σ donors. Crystallography showed a bridged chloride complex, with short strong bonds between the anion and both boron atoms.

Interestingly, *boromycin* (**3.10**), a naturally occurring antibiotic (Fig. 3.18) is structurally reliant on boron–anion interactions for its formation. The boron–oxygen donor atom interactions hold the structure of this molecule together, forming a tetrahedral anionic boron centre.

Other Lewis acids are more robust and have been used to build more structurally complex and selective anion receptors. Newcomb and co–workers have made extensive use of organotin macrocycles and cryptands for anion recognition (Fig. 3.19). Tetravalent organotin, although having a full valence shell of eight electrons, is still capable of accepting additional electron pairs into low–lying unoccupied molecular orbitals and consequently acting as a Lewis acid. Cryptands **3.11a** and **3.11b** were shown to encapsulate halide anions with 1:1 stoichiometry, and **3.11a** bound the fluoride ion ($K \approx 200$ M^{-1}) five times more strongly than chloride ($K \leq 0.01$ M^{-1}) in chloroform solution. Crystallography showed that the fluoride anion forms strong interactions with both tin atoms.

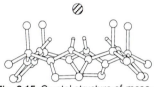

Fig. 3.15 Crystal structure of *meso*–octamethylcalix[4]pyrrole (**3.7**) with a bound chloride anion perched above the macrocyclic plane.

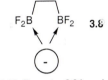

Fig. 3.16 Receptor **3.8** is a chelating ligand for anion binding

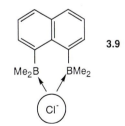

Fig. 3.17 Katz's rigid preorganized chelating ligand for anionic guests.

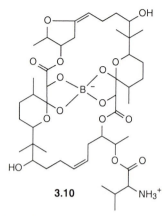

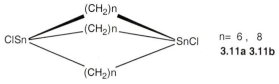

Fig. 3.19 Organotin cryptand encapsulates fluoride anions.

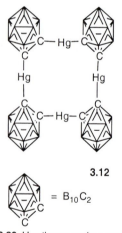

3.10

Fig. 3.18 Boromycin, a naturally occurring antibiotic, relies on boron anion interactions.

Organomercury receptors for anions have been of considerable interest. The mercury atom is sp hybridized and therefore contains two unfilled π orbitals into which it can readily accept electron pairs without any geometric reorganisation. In addition, it possesses a *linear geometry* and this has allowed the synthesis of large macrocyclic receptors. Hawthorne and his research group have assembled eyecatching receptors such as **3.12** (Fig. 3.20). This compound contains four carborane clusters, each of which contains ten boron atoms and two carbon atoms. These clusters are then bonded to mercury atoms through C–Hg–C linkers, leading to a macrocyclic structure. The formation of these macrocycles is templated by the presence of chloride anions (see Section 2.2 for an explanation of the template effect). Crystallography showed that the chloride anion binds in the plane of the macrocycle, equidistant from all four mercury atoms (Fig. 3.21). The iodide anion, however, is too large to fit into the cavity and two anions bind to the receptor, one above the plane of the ring, and one below it. This illustrates the importance of anionic radius for these receptors.

3.12

= B₁₀C₂

Fig. 3.20 Hawthorne and co–workers' mercuracarborand.

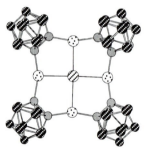

Fig. 3.21 Crystal structure of **3.12** showing chloride anion bound in the centre of the cavity.

All kinds of charged Lewis acidic species, such as Zn^{2+}, Cu^+ etc., can also be recruited for anion binding and are in fact frequently used by biological systems. Receptors for binding both positively and negatively charged guests together are, however, described in Chapter Four.

3.6 Recognition using combinations of interactions

As shown above, electrostatic interactions, hydrogen bonds and Lewis acids can each be used individually to bind anionic guests. Enhancements of both binding strength and selectivity can, however, often be obtained by *using intermolecular forces in concert*, rather than singly.

Polyaza macrocycles: electrostatics and hydrogen bonds

The strong binding afforded by electrostatic interactions can be enhanced by combination with hydrogen bonds. By chance, this powerful combination was actually used to create some of the very first anion hosts, which have since developed into one of the most important classes of anion receptor. In 1968, at around the same time that Charles Pedersen was reporting the crown ethers as hosts for alkali metal cations (Chapter Two), Simmons and Park reported a macrotricyclic receptor for anionic guests (**3.13**). The location of the bound halide anion was determined by nmr and, eventually, crystallographic methods. It binds in a *'katapinate'* manner; inside the macrotricyclic cavity between the two protonated nitrogen atoms, with the protons on these nitrogens oriented into the cavity, forming hydrogen bonds with the bound guest (Fig. 3.22). These hydrogen bonds complement the electrostatic forces making a strong complex.

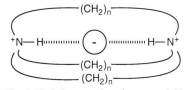

Fig. 3.22 Anion complex of receptor **3.13**.

Lehn and his co–workers, realized the importance of Simmons and Park's pioneering work, and developed a range of polyazamacrocyclic anion receptors (e.g. Fig. 3.23). They and others elucidated some key properties:

Electrostatic interactions are important: compound **3.14** binds chloride more strongly than iodide, because chloride is smaller and therefore presents a higher charge density to the host. This inherent electrostatic binding preference, however, can be *tuned by the receptor structure:* macrotricyclic **3.15**, for example, selects iodide over chloride.

Hydrogen bonds are also important: unlike macrocycles such as **3.14**, compound **3.16** does not bind highly charged ATP^{4-}, even though it possesses a high positive charge. This is because **3.16** cannot form hydrogen bonds with the anionic guest (Fig. 3.24).

These receptors are especially exciting because of their ability to bind *biologically relevant anions* (such as ATP^{4-}) in water. Receptor **3.17** provides a typical example (Fig. 3.25). The higher the charge on the phosphate anion, the stronger its recognition by the positively charged host, as a result of electrostatic interactions (Table 3.3). The receptor only operates efficiently at low pH, however, when all the nitrogen atoms are protonated. One way of extending operation to higher pH is to use longer alkyl spacers between the nitrogen atoms so that their positive charges do not repel one another so strongly. This stabilizes the protonated form of the host and means it exists and can bind anions in less acidic solutions (i.e. the pK_a's of the nitrogen atoms increase).

The binding sites of polyaza macrocycles can also be sterically *tuned*. Receptor **3.18** has two individual binding sites separated by alkyl chain spacers (Fig. 3.26). It forms the strongest complexes with dicarboxylate

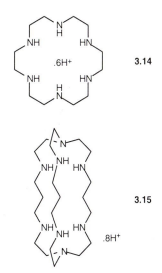

Fig. 3.23 Receptors for anions.

Fig. 3.24 Compound **3.16** does not bind ATP^{4-} as it can form no H–bonds

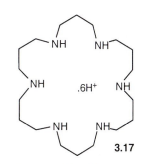

Fig. 3.25 Receptor **3.17** can bind biologically important phosphate guest anions.

Table 3.3 Binding constants (log K) for receptor **3.17** with biologically important guests in aqueous solution (0.1 M Me_4NCl).

Anion	log K
AMP^{2-}	3.4
ADP^{3-}	6.5
ATP^{4-}	8.9

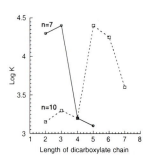

Fig. 3.27 **3.18a** (n=7) binds shorter chain dicarboxylates (x=2,3) more strongly, whilst **3.18b** (n=10) preferentially binds those with longer alkyl chains (x=5,6).

anions which have an alkyl chain of the correct length to fit within the cavity (Fig. 3.27). It shows good selectivity when compared with a simple macrocycle such as compound **3.14**.

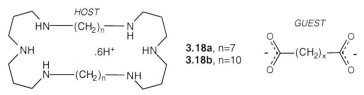

Fig. 3.26 Receptor **3.18** is specific for dicarboxylates of defined chain lengths.

As mentioned above, these receptors only operate at low to moderate pH when they are highly protonated. Many anions, such as phosphates, however, only become maximally negatively charged at higher pH values, when fully deprotonated. There is, therefore, a fine balance between protonation of the host and deprotonation of the anionic guest, which leads to a narrow pH zone in which the recognition of proton sensitive anions is possible (Fig. 3.28).

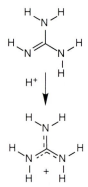

Scheme 3.2 Guanidine is readily protonated due to resonance stabilisation of the resultant cation (pK$_a$=13.6).

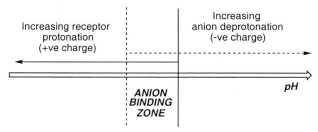

Fig. 3.28 Balance between host protonation and guest deprotonation.

Guanidinium: electrostatics and hydrogen bonds

Guanidine is readily protonated, forming guanidinium as the resultant cation, which is stabilized by resonance and charge delocalisation (Scheme 3.2). The pK$_a$ value is 13.6, making the guanidinium cation approximately three orders of magnitude more stable than a protonated secondary amine (pK$_a$≈10.5). Guanidinium therefore remains protonated up to high pH values, and is ideal for *extending the pH range over which anion receptors operate*. In fact, in biological systems, the arginine amino acid side chain has a terminal guanidinium group and this is extensively used for anion binding. Initial studies by Lehn and co–workers showed that anion recognition did indeed occur over a wide pH range There was, however, a price to pay for this significant gain: anion binding by macrocycles (e.g. **3.19**) was considerably weaker than by their polyammonium analogues, as a result of excess flexibility and charge delocalisation (Fig. 3.29). Nonetheless, guanidinium groups have proved versatile synthetic units for the construction of many anion receptors.

Schmidtchen and his research team overcame the flexibility of the acyclic guanidinium group by building the fragment into a *bicyclic ring* (Fig. 3.30). This harnesses another important feature of the guanidinium unit. The hydrogen bond donors, as for the urea based receptors discussed earlier

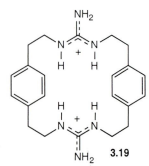

3.19

Fig. 3.29 A macrocyclic anion receptor containing guanidinium units

(Section 3.4), are suitably aligned for the recognition of *bidentate anionic guests*. This has led to extensive use of guanidinium based receptors for binding complementary carboxylate or phosphate guests. Receptor **3.20** (synthesized by de Mendoza, Lehn and co–workers) extracted a *complementary* guest (*p*–nitrobenzoate) from water into chloroform. In this case, the guanidinium–carboxylate interaction is further enhanced by π–π stacking, which provides additional selectivity (Fig. 3.31).

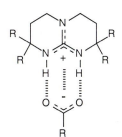

Fig. 3.30 Bicyclic guanidinium is preorganized and complementary for binding bidentate anions.

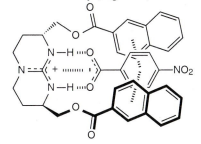

Fig. 3.31 Receptor **3.20** with a complementary bound anion.

Recently, guanidinium units have been built into a receptor which is very selective for citrate anions. This system will be discussed in Chapter Seven.

Expanded porphyrins: electrostatics and hydrogen bonds

A different approach to arranging protonated nitrogen atoms to construct anion receptors has been developed by Sessler and co–workers. They found that a protonated expanded porphyrin (sapphyrin) (**3.21**) that possesses a large electron poor cavity coordinates fluoride anions very strongly indeed (Fig. 3.32). Crystallography showed that the fluoride anion binds in the plane of the macrocyclic ring, and solution phase experiments indicate that it binds more strongly than bromide or chloride with a selectivity factor of over 10^3.

Sapphyrins have been functionalized in order to create more complex receptors (e.g. **3.22**) for biologically important anions such as anti–viral nucleic acid phosphates. Selectivity for guanine monophosphate is provided by the complementary hydrogen bonding triad (Fig. 3.33). This receptor selectively transports GMP through a model membrane system at neutral pH.

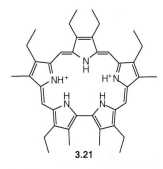

3.21

Fig. 3.32 Protonated sapphyrin (**3.21**) acts as a receptor for anionic guests.

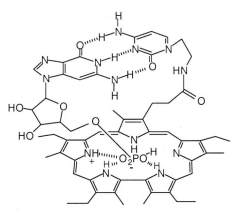

Fig. 3.33 Receptor **3.22** binds GMP in a complementary manner.

Fig. 3.34 Amide functionalized cobaltocenium binds anions.

Amide functionalized metallo compounds: electrostatics and hydrogen bonds

A different approach to using charged moieties and hydrogen bonding groups in order to snare anions has been employed by Beer and his research team. They have employed a *metal centre as a source of positive charge* and functionalized this with *secondary amide groups* capable of forming hydrogen bonds (e.g. Fig. 3.34). This general approach is analogous to many enzymes which have a combination of charged metal ions and hydrogen bonding groups at their recognition sites.

Cobaltocenium is a positively charged, 18 electron, organometallic sandwich complex and has been built into a range of receptors, which function in polar solvents such as acetonitrile and dimethylsulfoxide. Receptors **3.23** and **3.24** show how the stoichiometry of binding can be altered from 1:1 to 2:1 by using a long rigid spacer instead of a short flexible one (Fig. 3.35). These metallo-receptors are *readily tunable* and this means that *anion selectivities are consequently easily varied.*

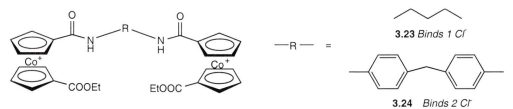

Fig. 3.35 Cobaltocenium based anion receptors have easily tunable binding sites.

Coordination compounds such as dipositive ruthenium bipyridine have also been functionalized with hydrogen bond donating amide groups. The crystal structure of receptor **3.25** (Fig. 3.36), in which a chloride anion is bound, clearly shows the binding site. The anion forms hydrogen bonds with the amide protons and also protons attached to the aromatic rings. As a consequence of the anion binding being proximate to the metal ion, these metallo–receptors can sense the presence of the bound guest through a change in the electrochemical or optical properties of the metal ion (see Chapter 7).

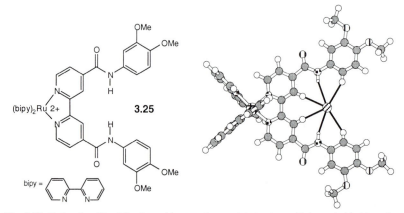

Fig. 3.36 Ruthenium bipyridine based host and a crystal structure with bound chloride anion.

Amide functionalized uranyl salenes: Lewis acid and hydrogen bonds

Lewis acids have also been assisted by hydrogen bonding interactions, notably in receptors such as **3.26** reported by Reinhoudt and co–workers (Fig. 3.37). This receptor contains a Lewis acidic uranyl salene group combined with hydrogen–bonding secondary amides. It binds the $H_2PO_4^-$ anion in polar organic solvents with stability constants as high as 10^5M^{-1}. Crystallography proved there was a strong bond between the uranium atom and an oxygen atom of the bound guest, with the secondary amides also participating in the binding, by forming hydrogen bonds.

Positively charged cyclophanes: electrostatics and hydrophobicity

Hydrophobic forces have also been used in combination with electrostatics to yield selectivity for different types of anionic guest (as expected from the Hofmeister series). Large *hydrophobic cavities* containing quaternized nitrogen atoms, such as **3.27** reported by Schneider and co–workers, have been synthesized (Fig. 3.38). This endows selectivity for *hydrophobic anionic guests*, such as nucleic acid base anions. In this way, the receptor uses its positive charge as a source of interaction with the anion, whilst the hydrophobicity provides the desired selectivity. This is similar to many biological anion binding sites which are buried in a hydrophobic pocket within the peptide superstructure.

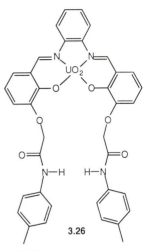

Fig. 3.37 Uranyl receptor for anionic guests combines Lewis acidity with hydrogen bonds.

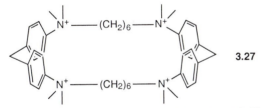

Fig. 3.38 A positively charged hydrophobic cavity binds hydrophobic anions.

Metallated cavities: electrostatics and hydrophobicity

Another approach to the combination of electrostatic and hydrophobic forces to trap anions has been taken by Atwood and his research group.

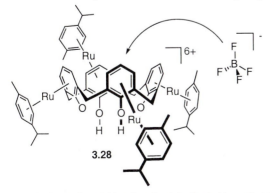

Fig. 3.39 A positively charged, metal functionalized, hydrophobic cavity snares anions.

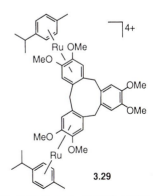

Fig. 3.40 Receptor for ReO_4^- anion.

They have functionalized the aromatic rings of *pre–existing, hydrophobic cavities*, such as calix[4]arenes, with *positively charged organometallic groups*. This yields electron deficient receptors such as **3.28** (Fig. 3.39), which has been shown by crystallography to encapsulate guests such as the tetrafluoroborate anion (BF_4^-) deep inside the cavity. More recently, a hydrophobic trigonal cyclotriveratrylene cavity has been functionalized with two organo–ruthenium groups (**3.29** in Fig. 3.40) and shown to bind the perrhenate anion (ReO_4^-). This is of special interest due to the close similarity between perrhenate and the environmentally problematic, radioactive pertechnetate anion.

3.7 Summary and conclusions

Anion recognition is a relatively new field of study which directly complements its more established sister field of cation coordination. Already, however, a dazzling range of anion receptors has been designed by supramolecular chemists, with much of the periodic table being utilized in their synthesis. Various binding forces have been manipulated, either singly, or in combination, in order to yield strong and sometimes selective recognition of anionic guests. As the factors controlling the strength and selectivity of anion coordination become better defined, these basic principles are now being used in the development of more complex supermolecules made up of multiple individual components (see Chapter Six).

Chemists working in the field are also beginning to turn to the development of functional devices based on the anion recognition event. This move towards function and applications is discussed further in Chapter Seven.

Suggested further reading

Anion binding is a recent development and consequently there are no introductory texts. There are, however, several worthwhile advanced texts.

The first book devoted to anion recognition is: *Supramolecular Chemistry of Anions*, Ed. A. Bianchi, K. Bowman–James, E. García–España, John Wiley and Sons, New York, 1997; in particular: Chapters 2, 3, 4 and 5. For a discussion of binding biologically relevant anions see: C. Seel, J. de Mendoza in *Comprehensive Supramolecular Chemistry*, Ed. J.L. Atwood, J.E.D. Davies, D.D. MacNicol, F. Vögtle, Vol. 2, Chapter 17, pp 519–552, Elsevier Science, Oxford, 1996. For a brief overview of early approaches to anion recognition see: B. Dietrich, *Pure and Applied Chemistry* **1993**, *65*, 1457–1464. For a review stressing the diversity of anion receptors see: K.T. Holman, J.L. Atwood, J.W. Steed in *Advances in Supramolecular* Chemistry, Ed. G.W. Gokel, Vol. 4, pp 287–330, JAI Press Inc., London, 1997; and for a discussion focussing on the variety of anion receptors based on inorganic chemistry see: P.D. Beer and D.K. Smith, *Progress in Inorganic Chemistry* **1997**, *46*, 1–96.

4 Simultaneous cation and anion binding

4.1 Introduction

The previous two chapters have shown the wide range of receptors for cationic or anionic guests which operate through a variety of intermolecular forces. There is, however, another approach to binding charged species, and that is to bind the cation and the anion with the same receptor. This type of recognition is referred to as *ditopic*. Three fundamentally different approaches to this goal have been employed; they will be outlined briefly here, and then discussed in further detail in the remainder of the Chapter.

The cascade approach

A receptor capable of binding more than one metal cation in well defined positions is synthesized. These 'recruited' metal ions are then used to to form interactions with an anionic guest, co-binding it between themselves and within the recognition cavity (Fig. 4.1). The overall complex is called a *cascade complex*, and models the way substrates are bound by many metallo-enzymes, for example alkaline phosphatase.

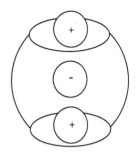

Fig. 4.1 A cascade complex.

Binding ion pairs using individual binding sites

Receptors containing two individual recognition units, one for a cation and one for an anion (Fig. 4.2), have also been synthesized. A question of crucial importance with such ditopic ion pair receptors is whether the binding of the cation enhances the binding of the anion, or vice versa (*cooperative recognition* through favourable electrostatics). There are many applications that could stem from selective ion pair recognition, for example, enhanced *extraction of toxic species* from aqueous industrial waste, enabling environmental clean-up.

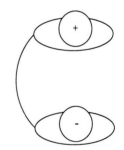

Fig. 4.2 Receptor with individual cation and anion recognition sites.

Binding zwitterions

Many important biological substrates, such as amino acids, contain both acidic and basic groups and consequently exist as zwitterions, with both a positive and a negative charge. For effective zwitterion recognition, the receptor must be designed such that the recognition units have the correct spatial arrangement to bind both charged parts of the molecule (Fig. 4.3). An important goal is the development of substrate selective receptors. The selective recognition of medically relevant zwitterionic guests under physiological conditions may lead to the development of new *sensors*.

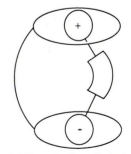

Fig. 4.3 Receptor for zwitterionic guests.

4.2 The cascade approach

In order to form a cascade complex the receptor must be capable of binding more than one metal ion. Some of the first such receptors were made by Robson and his co-workers in the early 1970s (Fig. 4.4). Compound **4.1** binds two metal ions, both of which are coordinatively saturated, and there is therefore no capacity to co-bind an anion. Compound **4.2**, however, binds *coordinatively unsaturated* metal ions, and therefore has the ability to co-bind a small anion such as an alkoxide (X) between the two metals.

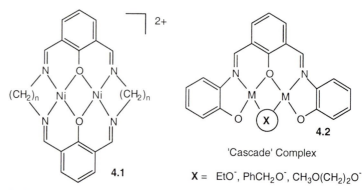

Fig. 4.4 Receptor **4.1** has no space for substrate binding, whilst receptor **4.2** contains a pair of coordinatively unsaturated metals and can therefore form cascade complexes.

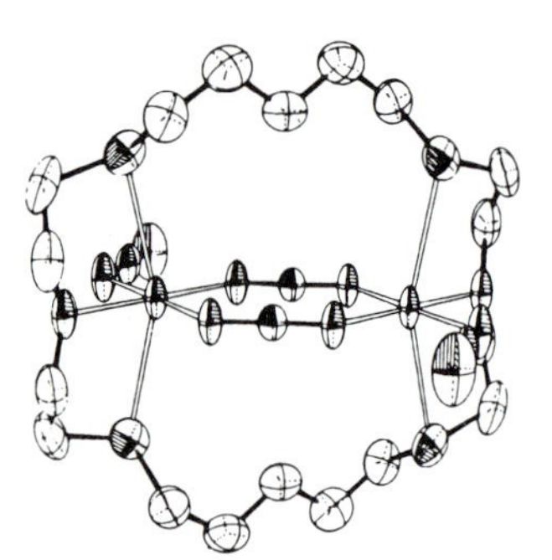

It was Jean-Marie Lehn who highlighted the importance of cascade systems to the developing field of supramolecular chemistry. His research group then synthesized a range of receptors suitable for cascade complex formation.

Fig. 4.5 Two azide anions bridge the two metal centres of receptor **4.3** forming a cascade complex. There is also one azide anion bound in a terminal mode to each metal ion. Reprinted with permission from J. Am. Chem. Soc. 1979, 101, 3381. Copyright (1979) American Chemical Society.

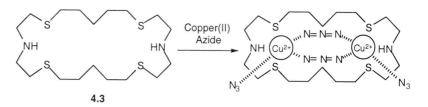

Scheme 4.1 Macrocyclic receptor **4.3** forms a cascade complex with copper(II) azide.

Receptor **4.3** (Scheme 4.1) can bind two copper(II) ions and has an ellipsoidal shape, with the metal ions held relatively distant from one another. Ellipsoidal azide guest anions can therefore bridge these metal ions, whereas the smaller, spherical chloride anion coordinates to each metal ion individually (Figs. 4.5 and 4.6). Receptor **4.3** illustrates how a macrocyclic host can control the spacing of metal ions and manipulate the selectivity of cascade complex formation. This cascade binding motif is analogous to the way in which some copper based enzymes bind azide anions. As well as using macrocyclic ligands for the formation of cascade complexes, cryptands have been used to enhance the strength and selectivity of binding.

Recently, more rigid cavities have also been used in order to organize arrays of metal ions. A bucket shaped calix-resorcinarene cavity has been used

Fig. 4.6 Chloride anions are too small to bridge between the metal ions so two anions coordinate to each metal cation forming a traditional coordination complex. Reprinted with permission from J. Am. Chem. Soc. 1979, 101, 3381. Copyright (1979) American Chemical Society.

to organize an array of donor phosphorus atoms around the upper rim (Fig. 4.7). This ligand is capable of binding four metal ions (either Cu(I) or Ag(I)), and these in turn can imprison an anion in a cascade manner (Scheme 4.2).

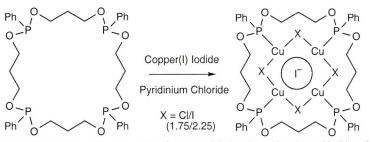

Scheme 4.2 View from above the cavity of receptor **4.4** showing an imprisoned iodide anion.

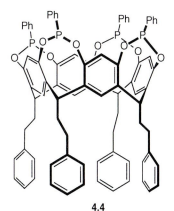

4.4

Fig. 4.7 Bucket shaped receptor for the formation of cascade complexes.

Receptor **4.4** was treated with copper(I) iodide and pyridinium chloride in order to discover which anion was selected (Scheme 4.2). Interestingly, the four anions (X) around the rim of the receptor were either iodide (2.25 on average) or chloride (1.75 on average), but the iodide anion was exclusively bound in the centre of the cavity. This size selective binding mode was proven by crystallography (Fig. 4.8). In addition, the nucleophilic properties of the imprisoned anion were dramatically changed. Increasing the number of metal ions and arranging them around a pre-formed cavity therefore yields strong and selective anion binding.

Receptor **4.5** co-binds phosphate anions. If the phosphate anion was a suitable phosphodiester then the receptor catalysed its intramolecular hydrolysis (Fig. 4.9). When the catalytic ability of receptor **4.5** was compared with a simple analogue capable of binding only one copper(II) ion a rate enhancement was observed. Many enzymes use metal ions in this way to recruit substrate anions prior to catalytic transformation. This indicates the ability of cascade complexes to model not only the recognition properties of enzymes such as phosphatases, but also their catalytic action

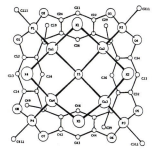

Fig. 4.8 Crystal structure of top face of receptor **4.4** with coordinated copper iodide. Reprinted with permission from J. Am. Chem. Soc. 1995, 117, 8362. Copyright (1991) Ameican Chemical Society.

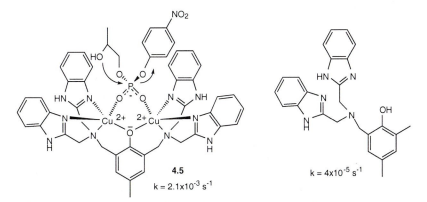

Fig. 4.9 Complex **4.5** catalyses intramolecular hydrolysis of phosphodiesters more efficiently than a mononuclear analogue.

A cascade type complex using non-coordinate interactions has also been reported. Receptor **4.6** binds two alkali metal cations, one in each crown ether ring (Fig. 4.10). These electrostatically bound cations are subsequently capable of co-binding an anionic substrate *via* electrostatic attraction to their own positive charges. In addition, the binaphthyl unit possesses chirality due to its twist away from planarity, and the receptor with potassium ions bound in the crowns exhibited a 15% enantiomeric excess in the recognition of optically active mandelate anions (Fig. 4.11).

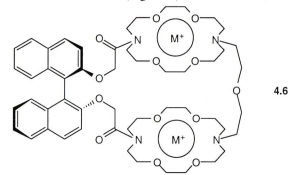

4.6

Fig. 4.11 Optically active mandelate guest anion.

Fig. 4.10 Receptor for electrostatic cascade complex formation (M⁺ = alkali metal cation).

4.3 Binding ion pairs with individual binding sites

The second approach to binding ion pairs involves designing a receptor with both cation and anion binidng sub-units. In Chapter Three, compound **3.3** was introduced as an anion receptor operating through dipole-anion interactions. This host can also bind cations using its electronegative oxygen atoms. In this case the binding of the cation and anion act in opposition to one another and the binding is anti-cooperative (i.e cation binding does not enhance the strength of anion recognition or vice versa).

Table 4.1 Salt binding characteristics of receptor **4.7**.

Salt	Cation Bound?	Anion Bound?
KF	✔	✔
KCN	✔	✔
KOMe	✔	✔
KI	✔	✘
KSCN	✔	✘
KCl	✘	✘
KBr	✘	✘

4.7

Fig. 4.12 Ditopic receptor for binding ion pairs.

Cooperative ditopic receptors for ion pairs have, however, also been reported. This field of study was largely established by Manfred Reetz and his research team, who synthesized receptors such as **4.7** (Fig. 4.12). This molecule contains both a crown ether unit (cation binding site) and a Lewis acidic boron atom (anion binding site). Receptor **4.7** was capable of dissolving anhydrous KF in dichloromethane solution. ¹¹B nmr proved anion binding, whilst ¹H and ¹³C nmr showed incorporation of the potassium ion in the crown. A crystal structure confirmed this mode of binding. The positive effect of covalently linking the two binding sites was proven, as the ditopic receptor was more efficient than a simple 1:1 mixture of crown ether and boronic acid. The ion pair recognition is therefore cooperative. The recognition of other potassium salts was also investigated (Table 4.1).

Receptor **4.8**, synthesized by Reinhoudt and co-workers, displays very elegant cation–anion cooperativity. The receptor consists of a calix[4]arene with cation binding ester groups at the lower rim and anion binding urea groups at the upper rim. In chloroform solution the two urea groups are hydrogen bonded together, and so are not available for hydrogen bonding to any putative anionic guest species. However when sodium cations are added, they bind to the ester groups of the calixarene, causing the lower rim to contract. This forces apart the urea groups at the upper rim apart which are then available for binding anionic species such as chloride (Scheme 4.3).

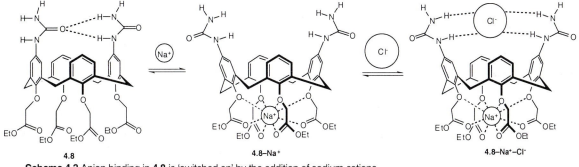

Scheme 4.3 Anion binding in **4.8** is 'switched on' by the addition of sodium cations.

4.4 Binding zwitterions

Receptor **4.9** links an aza crown (cation binding) with a macrotricyclic anion receptor (**3.1a**) and has been used for the recognition of zwitterionic guests (Fig. 4.13). Both parts of the receptor were shown to be important for binding. There was, however, little selectivity between different zwitterions, presumably because the 1,4-dimethylbenzene spacer group is flexible, which means the distance between the two recognition units can easily vary (Fig. 4.14).

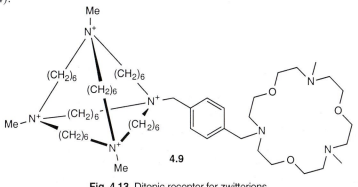

Fig. 4.13 Ditopic receptor for zwitterions.

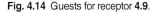

Fig. 4.14 Guests for receptor **4.9**.

Zwitterion recognition is of great biological importance, as it is essential for amino acid binding. Receptor **4.10** provides a most elegant example of biologically relevant zwitterion recognition. It extracts amino acids from aqueous solution into dichloromethane. The extraction occurred selectively, with a strong preference for amino acids with aromatic side chains, such as phenylalanine and tryptophan. A three point binding mode was proposed (Scheme 4.4). The carboxylate anion is bound by the guanidinium unit, the protonated amine by the aza crown, and π–π interactions between the phenyl side chain of the amino acid and the receptor's naphthalene unit enhance the strength and selectivity of complex formation. What is more, receptor **4.10**, which is optically active, was completely selective for L-amino acids over D-amino acids, with no extraction of D-amino acids being observed. This indicates the excellent design and spatial arrangement of the subunits in the receptor.

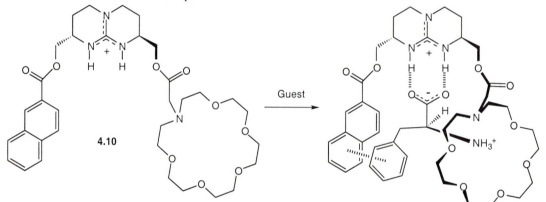

Scheme 4.4 Three point receptor (**4.10**) for the selective recognition and extraction of zwitterionic amino acids.

4.5 Summary and perspectives for the future

The simultaneous recognition of cationic and anionic guests is a very young field of research and currently extremely vigorous. It is expected that in the coming years, as the individual research areas of cation and anion coordination become more sophisticated, research into binding ion pairs and zwitterions will catch up with, and probably even extend, the state of the art.

In particular, it is expected that biologically important guests will be bound with ever increasing selectivity and that devices suitable for industrial or medical use will be forthcoming.

Suggested further reading

For an overview of cascade complexes see: J-M. Lehn, *Pure and Applied Chemistry*, **1980**, *52*, 2441-2459. For a review of ion pair receptors see: M.T. Reetz in *Comprehensive Supramolecular Chemistry*, Ed. J.L. Atwood, J.E.D. Davies, D.D. MacNicol, F. Vögtle, Vol. 2, Chapter 18, pp 553-562, Elsevier Science, Oxford, 1996. The reader is also strongly recommended to the excellent research article about the enantioselective recognition of amino acid zwitterions: A. Galán, D. Andreu, A.M. Echavarren, P. Prados, J. de Mendoza, *Journal of the American Chemical Society* **1992**, *114*, 1511-1512.

5 Neutral guest binding

The previous chapters have focused on the recognition of ionic guests. This chapter, however, discusses the recognition of uncharged guests, a goal which requires the use of different strategies to achieve strong, selective recognition. Neutral guest binding is a difficult task — the armoury of non-covalent interactions at our disposal now includes hydrogen bonding, π–π stacking, the hydrophobic effect and charge transfer interactions. These interactions are highlighted in the examples illustrated in this chapter.

5.1 Hydrogen bond receptors

Biological relevance

Hydrogen bonds are essential in biological systems. Their *directionality* means that host and guest must be perfectly aligned for interaction to occur and hydrogen bonds therefore help enzymes acquire the selectivity they require. An example of the importance which a *single hydrogen bond* can possess is provided by vancomycin, the current antibiotic of last resort. Vancomycin operates by forming a hydrogen bonded complex with a peptide that is terminated by two D-alanine amino acids (D-Ala-D-Ala terminus) (**5.1**). This peptide building block is required by bacteria for their cell wall synthesis. Vancomycin binding prevents cell wall synthesis, killing the bacteria. Bacteria resistant to this powerful antibiotic are, however, beginning to emerge. They have evolved to synthesise their cell walls using peptides terminated with D-Ala-D-lactate (**5.2**). In other words, one amide group is replaced by an ester. This removes a single hydrogen bond from the complex, and prevents vancomycin from destroying the bacteria (Fig. 5.1).

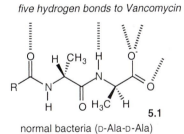

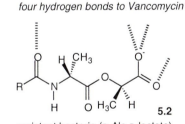

five hydrogen bonds to Vancomycin *four hydrogen bonds to Vancomycin*

5.1 **5.2**

normal bacteria (D-Ala-D-Ala) resistant bacteria (D-Ala-D-lactate)

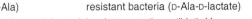

Fig. 5.1 A single hydrogen bond controls bacterial resistance to the antibiotic Vancomycin.

The exquisite control which hydrogen bonds can yield has appealed to supramolecular chemists who want to direct their molecular building blocks as precisely as possible. In addition, many *biologically important guest molecules* have multiple hydrogen bonding sites.

Complementary hydrogen bonding receptors

A well designed hydrogen bonding receptor should possess hydrogen bonding groups which are *complementary* to the desired guest. These groups must be *preorganised* and *rigidly held*, so that they will interact with the guest molecule rather than collapsing down onto each other. Receptor **5.3**, synthesised by Andrew Hamilton and co-workers forms six hydrogen bonds with a complementary guest, such as a barbiturate in non-competitive solvents (Fig. 5.2). Barbiturates are particularly interesting substrates due to their medical use as sedative and anti-convulsant drugs. The binding constant is highly dependent on the exact structure of the guest indicating the importance of good complementarity (Table 5.1). The strongest binding is observed when all six hydrogen bonds can be formed. When two of the carbonyl oxygens are removed, the complex formed has two less hydrogen bonds and, as a consequence, is much weaker. A crystal structure of the complex was obtained (Fig. 5.3), providing further evidence for the binding mode.

Table. 5.1 Binding data for receptor **5.3**.

Guest	K (M⁻¹) [CH₂Cl₂]
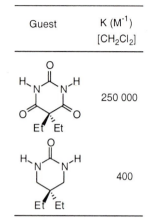	250 000
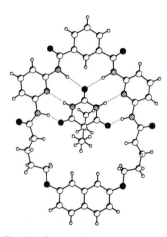	400

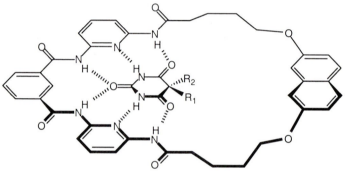

Fig. 5.2 Host-guest complex between receptor **5.3** and a barbiturate guest.

Receptor **5.4** satisfies every possible hydrogen bond requirement of its complementary guest, urea (Fig. 5.4). It is based on a fused ring system, which provides a high degree of rigidity and consequently, *preorganisation*. The recognition of urea was remarkably strong, with the complex even surviving in hot dimethylsulfoxide (this solvent is an excellent hydrogen bond acceptor and usually disrupts neutral hydrogen bond receptors). Receptor **5.4** illustrates how, with careful design, hydrogen bond recognition can yield a strong complex even in relatively competitive solvent media.

Fig. 5.3 Crystal structure of receptor **5.3** with bound barbiturate guest. Reprinted with permission from J. Am. Chem. Soc. 1991, 113, 7640, Copyright (1991) American Chemical Society.

Secondary interactions

Using arrays of hydrogen bonds, however, is not always as straightforward as choosing a complementary host-guest pair. Consider, for example, receptor **5.5**: it possesses Donor-Acceptor-Donor (DAD) hydrogen bonding fragments complementary to the Acceptor-Donor-Acceptor (ADA) groups of the guest (Fig. 5.5). Hydrogen bonds have considerable *electrostatic character*, being formed between an electropositive hydrogen atom and an electronegative heteroatom. As Jorgensen first pointed out, this electrostatic character means

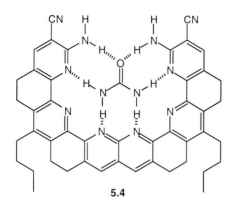

Fig. 5.4 Receptor for urea which satisfies every hydrogen-bond requirement of the guest.

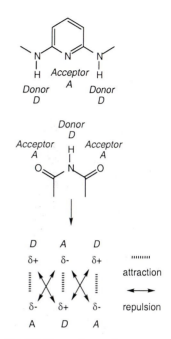

that *diagonal interactions* in the array are also important in determining binding strength, and he referred to them as *secondary electrostatic interactions*. In an H-bond array, centres of like charge repel one another diagonally, weakening the binding (Fig. 5.5). These secondary interactions can also be constructive, with the donors and acceptors organised such that all interactions are attractive, strengthening the array (Fig. 5.6). Calculations indicated that diagonal secondary interactions contribute approximately one third of the strength of the primary hydrogen bonds themselves.

Fig. 5.5 Diagonal secondary electrostatic interactions modulate the strength of interaction of an H-bonding DAD•ADA array.

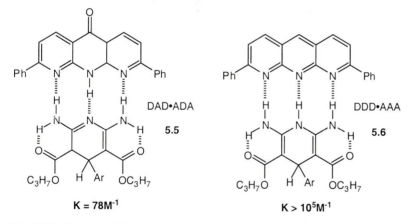

Fig. 5.7 Binding strength depends on the order of donors and acceptors in the H-bond array.

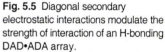

Fig. 5.6 Totally constructive diagonal secondary interactions in a DDD•AAA array.

The secondary interaction hypothesis was experimentally proven by elegant work from the research group of Steve Zimmerman. The strength of interaction of a range of hosts and guests with different donor-acceptor arrays was investigated. The arrangement of donors and acceptors does indeed play a critical role in controlling the strength of complex formation. This is clearly illustrated by complexes **5.5** and **5.6**, which, although both containing three hydrogen bonds, have vastly different formation constants (Fig. 5.7). The DAD•ADA array (**5.5**) contains only *destructive* secondary interactions and is weakly held, whilst DDD•AAA (**5.6**) has only *constructive* secondary interactions and is consequently favoured. Complexes of the type DDA•AAD have intermediate binding strengths as two of the secondary interactions are attractive and two repulsive.

Chiral recognition

In previous chapters we have introduced optically active receptors, which are capable of chiral recognition. This is a research target of particular importance because many biologically and medically important molecules exist as enantiomers. The directionality and controllability of hydrogen bonds makes them ideal for achieving chiral discrimination.

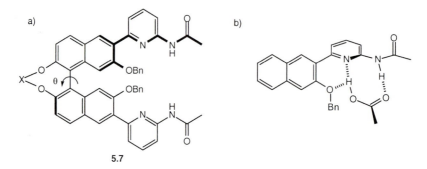

Fig. 5.8 a) The conformation of chiral cleft receptor (**5.7**) is dependent on the spacer group (X) b) Receptor **5.7** has two recognition sites for carboxylic acid groups.

Table 5.2 Receptor **5.7** binds one enantiomer of N-protected aspartic acid more strongly than the other. The enantioselectivity is tuned by the spacer group (X) which controls the angle (θ).

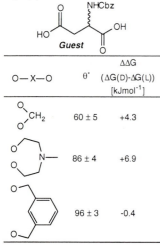

O—X—O	θ°	ΔΔG (ΔG(D)-ΔG(L)) [kJmol⁻¹]
CH₂	60 ± 5	+4.3
N—	86 ± 4	+6.9
	96 ± 3	-0.4

Diederich and his research group at ETH Zürich have synthesised receptors for amino acid recognition. Receptor **5.7**, for example, provides a cleft environment suitable for binding a dicarboxylic acid (Fig. 5.8). In addition, the twisted binaphthalene unit has a chirality axis, placing the binding site in a chiral environment. Protected L(and D)-aspartic acid (a neurotransmitter) was bound in non-competitive chloroform solution. The enantiomeric receptor bound the two enantiomeric guests with different strengths (Table 5.2). The difference in binding strength for the two enantiomers (the enantioselectivity, ΔΔG) was as high as 6.9 kJmol⁻¹. The magnitude of this enantioselectivity was dependent on the spacer group (X), which alters the dihedral angle (θ), and consequently the relative orientation of the receptor's hydrogen bonding groups (Table 5.2). This shows that by tuning the highly directional hydrogen bonding environment, the enantioselectivity can be enhanced.

Sugars are also chiral biological molecules of great importance. They possess a complex, three dimensional skeleton, however, and it is therefore difficult to design a perfect complementary host. Typical approaches to the goal of selective sugar binding have focussed on creating a binding cleft or cavity, such as receptor **5.8**, which contains a number of convergently focussed hydrogen bonding groups. This macrocyclic host is built from two naturally occurring, cheap, chiral steroid building blocks. It discriminates between α- and β- substituted diastereomeric sugars as well as the L- and D-enantiomers (Fig. 5.9).

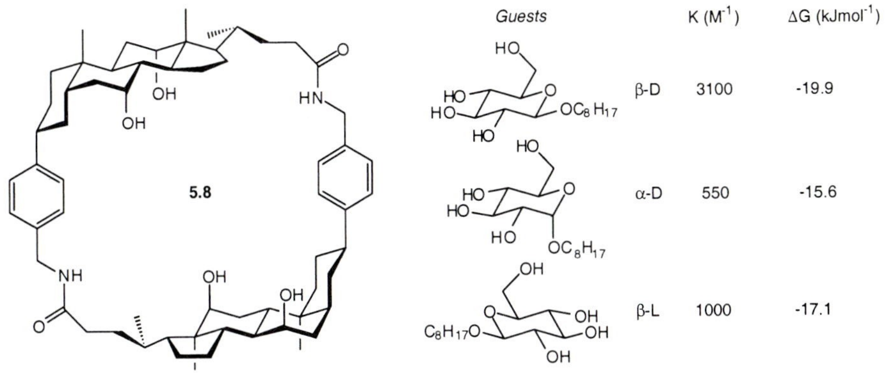
5.8
Guests
K (M^{-1})
ΔG ($kJmol^{-1}$)
β-D 3100 -19.9
α-D 550 -15.6
β-L 1000 -17.1

Fig. 5.9 Macrocyclic receptor for sugar guests shows both diastereo- and enantio-selective recognition.

Supplementing hydrogen bonds with other interactions

Biological systems, however, are not solely dependent on hydrogen bonds for molecular recognition. Often, hydrogen bonds provide *organisation* and *selectivity*, but their strength is enhanced by the use of other intermolecular interactions; in particular, *π–π interactions* and *hydrophobic effects*. Combining interactions in this way can lead to new modes of binding, as well as enhanced binding selectivity and strength.

Andrew Hamilton and co-workers reported a beautiful example of π–π stacking enhancing hydrogen-bond mediated recognition. Receptor **5.9** contains a DAD hydrogen bonding array, suitable for binding thymine derivatives (thymine is one of the four DNA bases). The receptor, however, also contains a naphthalene unit attached by flexible spacers, which can π–π stack with the aromatic ring of the bound substrate. Both nmr and crystallography illustrated that the receptor acts as a hinge, with the aromatic ring forming additional interactions with the bound guest (Scheme 5.1).

Julius Rebek and his research team synthesised *molecular cleft* **5.10** which has *convergent* hydrogen bonding groups (Scheme 5.2). These groups are complementary to derivatives of the nucleobase adenine, and the rigidity of the cleft prevents the hydrogen bond donors and acceptors from collapsing. Additionally the naphthalene spacer group of the receptor can stack with the π-system of the adenine guest species.

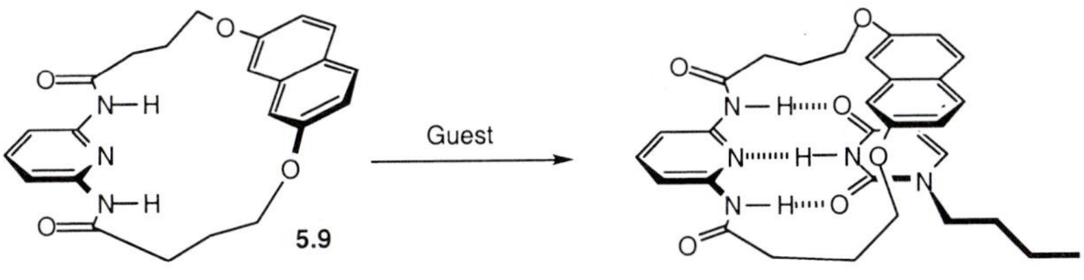
5.9
Guest

Scheme 5.1 Molecular hinge for the complexation of complementary nucleoside bases.

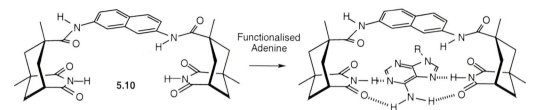

Scheme 5.2 Receptor **5.10** has a rigid cleft with preorganised hydrogen bonding groups complementary for binding adenine derivatives.

Receptor **5.11**, reported by Christopher Hunter, James McQuillan and their co-workers binds benzoquinone *via* the formation of hydrogen bonds, enhanced by π–π interactions (Fig. 5.10). What is more, the π–π interactions do not merely assist in binding; they act cooperatively with the hydrogen bonds to polarise the guest, altering its electronic properties. These properties were monitored using electrochemistry, and UV-visible spectroscopy. Their modulation clearly illustrates how molecular recognition can alter the *physical properties* of a bound guest.

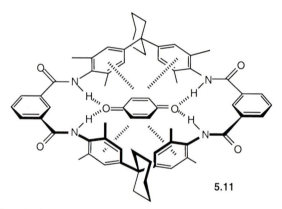

Fig. 5.10 Receptor for benzoquinone which alters the electronic properties of the guest.

The receptors discussed thus far operate in *non-competitive solvents* (e.g chloroform) which do not disrupt recognition. It is ultimately desirable, however, to bind biologically important guests in competitive solvents such as water (the biological medium). The use of the hydrophobic effect is a common feature in receptors that achieve this goal.

5.2 The hydrophobic effect: recognition in water

Cyclophanes as receptors for apolar guests

Perhaps the most obvious way of harnessing hydrophobicity is to build a macrocycle containing 'greasy' walls with a cavity large enough to encapsulate suitable hydrophobic guests.

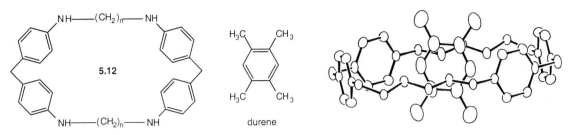

Fig. 5.11 Cyclophane **5.12** binds apolar guests such as 1,2,4,5-tetramethylbenzene (durene). Crystal structure reprinted with permission from J. Am. Chem. Soc., 1980, 102, 2504, Copyright (1980) American Chemical Society.

A *cyclophane* is a cyclic molecule which possesses at least one aromatic ring bridged by aliphatic chains. In order to bind a guest within the cyclophane using hydrophobic interactions, we are particularly interested in cyclophanes which have *parallel aromatic walls* which create a *well defined cavity*. This class of molecule has been known for many years, and the potential for guest encapsulation was considered as early as 1955, but most of the systems available at this time were too small. The first unambiguous example of inclusion within a cyclophane cavity (**5.12**) was only obtained in 1980 by Koga and co-workers (Fig. 5.11). The amines within the cyclophane cavity mean that the receptor is protonated at low pH, and consequently soluble in acidic aqueous solution. Binding studies were therefore carried out at pH<2. ^{1}H NMR was particularly useful in providing information about the inclusion complexes, for example the signals of bound 2,7-dihydroxynaphthalene shifted dramatically (Fig. 5.12). An acyclic analogue exhibited no recognition, showing the importance of having a preorganised cavity. Crystallography finally provided convincing evidence of the binding mode with a crystalline complex of stoichiometry **5.12**.durene.4HCl.4H$_2$O being obtained (durene is 1,2,4,5-tetramethylbenzene). Durene was deeply included within the cyclophane cavity (Fig. 5.11). 1-Anilinonaphthalene-8-sulfonate was also used as a guest, its fluorescence spectrum was enhanced in the presence of the cyclophane. The fluorescence data was used to generate a binding constant (6250 M^{-1}), illustrating the strong binding which can be provided by hydrophobicity.

The aromatic groups in the walls of the cyclophane endow the receptor with two important features. Firstly, they are rigid, and consequently, the cyclophane possesses a well defined, preorganised cavity, suitable for guest binding. Secondly, they are hydrophobic groups and provide the thermodynamic strength for recognition.

After this initial report of hydrophobic guest binding in aqueous solution, further cyclophanes such as **5.13** followed (Fig. 5.13). The cavity in this receptor is narrow and flat, and is ideally suited for the incorporation of flat aromatic guests. Much use has been made of this cyclophane for the assembly of more complex molecular architectures *via* charge transfer interactions, as will be illustrated in Chapter Six.

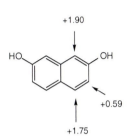

Fig. 5.12 ^{1}H NMR peak perturbations when **5.12** binds 2,7-dihydroxynaphthalene.

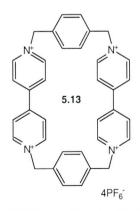

Fig. 5.13 Paraquat based cyclophane with flat binding cavity suitable for aromatic guests.

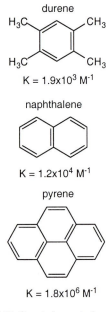

durene

H₃C, CH₃

H₃C CH₃

K = 1.9x10³ M⁻¹

naphthalene

K = 1.2x10⁴ M⁻¹

pyrene

K = 1.8x10⁶ M⁻¹

Fig. 5.15 Guests for cyclophane **5.14**

Cyclophanes **5.12** and **5.13** are solubilised by charged groups in the walls of the cyclophane cavity. These hydrophilic groups, however, reduce the degree of hydrophobic stabilisation that occurs on guest binding. If the charges required for solubilisation are held outside the macrocyclic ring they cannot interfere with either the strength or the selectivity of binding. The first cyclophanes with *external charges* were reported by Diederich and co-workers (Fig. 5.14). Crystallography indicated that their cavities were large and open, preorganised for the recognition of hydrophobic guests. Receptor **5.14** can solubilise or extract polycyclic aromatic hydrocarbons (PAHs) into aqueous solution and binding constants were high. In particular, guests with larger hydrophobic surfaces (such as pyrene) bind more strongly than smaller guests (such as durene or naphthalene) due to better stereoelectronic complementarity (Fig. 5.15). Receptor **5.14** enhances the transport rate of pyrene through an aqueous phase 430-fold (but durene only by a factor of 2).

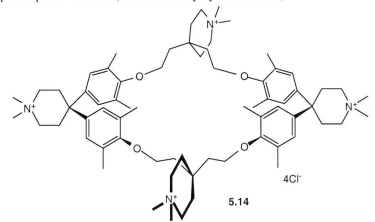

4Cl⁻

5.14

Fig. 5.14 Receptor with charge solubilisation outside the macrocyclic cavity.

Effect of solvent

Thus far, we have simply stated that cyclophanes bind guests through hydrophobic interactions with apolar guests in water. In a key study, Diederich investigated the exact role of the solvent in this type of recognition.

The binding constant of a cyclophane-type receptor with pyrene was monitored in solvents ranging from very polar H_2O/DMSO (99/1) to apolar carbon disulfide (CS_2). The binding was strongest in water, and generally decreased as the polarity of the solvent decreased. Binding still occurs in non-aqueous solvents, however, and the binding strength in cyclophanes can therefore be considered to come from a generalised 'solvatophobic' effect. Two main conclusions about the role of the solvent were drawn from this study: a) if the solvent molecules interact strongly with one another (*high cohesive interaction strength*), then complexation is stronger, because the equilibrium in Fig. 1.17 (see Chapter One) is pulled to the right; b) if the solvent has *high polarizability*, then complexation is weaker as the solvent interacts efficiently with the uncomplexed host and guest through van der Waals interactions, pulling the equilibrium to the left. In fact, the binding strength was directly correlated with an empirical solvent parameter (E_T) which incorporates these two factors. These observations explain why water is such an ideal solvent for apolar recognition using cyclophanes. It has strong cohesive interactions due to its ability to form extended hydrogen bond networks, and its polarizability is very low. Consequently, guests bind very strongly in water, and the source of binding strength is specifically referred to as the 'hydrophobic' effect.

Inclusion within other receptor cavities: cyclodextrins

Cyclodextrins are naturally occurring macrocycles composed of sugar units and were first reported as long ago as 1891. There was at this stage, however, no inkling of their structure or potential properties. They are obtained by the enzymatic degradation of starch, a process which yields linear and cyclic oligomers of glucose (referred to as dextrins and cyclodextrins respectively). Their purification was achieved in the 1950s, and by the end of the 1960s, concurrent with the rise of supramolecular chemistry, their potential to act as host molecules was being actively investigated. There are three particularly important cyclodextrins: α-Cyclodextrin contains 6 D-glucose units, β-cyclodextrin (**5.15**) 7, and γ-cyclodextrin 8 (Fig. 5.16 and Table 5.3). One rim of the cavity possesses secondary hydroxyl groups, whilst the other rim is functionalised with primary hydroxyl groups. The interior of the cavity is non-polar and its ability to bind different guests has been of great interest.

Initially, the ability of unfunctionalised cyclodextrins to act as hosts was investigated with a range of different guests, such as steroids. Steric and electronic properties of the guest both play important roles. The primary source of binding strength is a matter of some debate, but the hydrophobic effect is believed to play a role. Hydrogen bonds or charge stabilisation from the large number of OH groups can, however, also contribute to the binding strength and alter the selectivity. Additionally, they permit the coordination of more polar guests. The hydroxyl groups also allow functionalisation of the cyclodextrin cavity and many reports have been made, often with the goal of sensing the presence of the bound guest or enhancing catalysis (see Chapter 7). Cyclodextrin chemistry is a huge and active area of research, primarily due to the industrial availability of the macrocycles and their extremely low cost.

Inclusion within other receptor cavities: calixarenes

The cation coordination abilities of calix[4]arenes were discussed in Chapter Two. Calix[n]arenes, however, are also a type of cyclophane possessing a hydrophobic cavity with aromatic walls. They can be synthesised with different ring sizes, the commonest being those containing 4, 6 or 8 aromatic rings (calix[4]arene, calix[6]arene and calix[8]arene respectively). Calix[4]arenes such as *p-tert*-butylcalix[4]arene have been shown to bind small neutral guests such as toluene in both the solid and solution phase. The crystal structure indicated that the aromatic molecule was bound within the calix[4]arene cavity *via* CH-π hydrogen bonds (Fig. 5.17). Solution phase studies indicated very weak toluene binding (K= 1.1 M^{-1}) but as the recognition was investigated in chloroform, in which no strong solvatophobic effect can operate, this is not surprising.

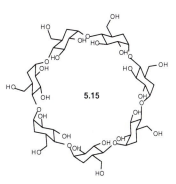

Fig. 5.16 β-Cyclodextrin.

Table 5.3 Cavity sizes of the different cyclodextrins.

Cyclodextrin	Cavity Diameter	Cavity Depth
α	5.7Å	7.8Å
β	7.8Å	7.8Å
γ	9.5Å	7.8Å

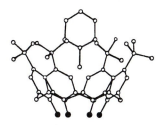

Fig. 5.17 *p-tert*-Butylcalix[4]arene-toluene inclusion complex (reproduced with permission from J. Chem. Soc. Chem. Commun. 1979, 1005. Copyright (1979) The Royal Society of Chemistry).

As the calixarene ring becomes bigger, the potential to bind larger, more interesting guests increases. In particular, calixarenes have been used to bind fullerenes, a topic which will be discussed further in Chapter Seven.

Inclusion within other receptor cavities: carcerands

Donald Cram and his research team have specialised in the synthesis of unique functional molecular architectures, and as part of this program they developed the *carcerands* (**5.16**). These receptors are structurally related to the calixarenes and contain concave internal surfaces suitable for guest recognition. They can be visualised as two bowls which have been covalently connected in a 'head to head' manner forming a molecular capsule (Fig. 5.18).

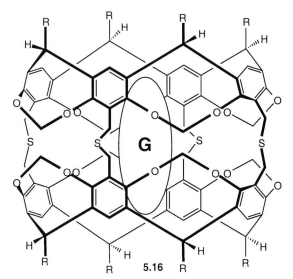

Fig. 5.18 Carcerand receptor acts as a molecular cage, imprisoning a guest molecule **G**.

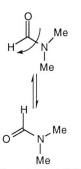

Fig. 5.19 The rotation of DMF trapped inside a carcerand cage has been investigated.

During carcerand synthesis, small molecules such as ethanol, dimethyl-formamide (DMF), acetonitrile etc., became *trapped* inside the host. They can interact with the receptor through *van der Waals forces*, and once synthesis is complete, they cannot exit from the sealed capsule for steric reasons. They can be considered as 'incarcerated', and such complexes are referred to as carceplexes. Carcerands can be used to alter the properties of the imprisoned guest, for example, the rotational energy barrier of DMF is lower in a carceplex than when uncomplexed (Fig. 5.19).

Such receptors, however, do not allow recognition to occur once they have been synthesised, as the cavity is completely inaccessible. This led Cram and his co-workers to synthesise *hemi-carcerands*, in which some of the spacer groups holding the hemispheres together are absent, or replaced with larger groups. These receptors then possess *larger entry and exit portals* around the recognition site and allow slow exchange of guest molecules (heat is required to force the guest out of the cavity entropically). These hemi-carcerands show a range of interesting recognition properties, including the stabilisation of otherwise unstable guest molecules, such as cyclobutadiene (Fig. 5.20).

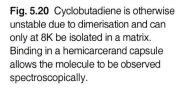

Fig. 5.20 Cyclobutadiene is otherwise unstable due to dimerisation and can only at 8K be isolated in a matrix. Binding in a hemicarcerand capsule allows the molecule to be observed spectroscopically.

5.3 Coordination through dative bond formation

Coordination chemistry or supramolecular chemistry?
When any new field of chemistry develops, there is naturally some overlap with other well established areas. Classical coordination chemistry has had a strong influence on supramolecular chemistry. In coordination chemistry, the formation of strong dative bonds enables guest binding even in competitive media. Many examples of the use of metal complexes, such as metalloporphyrins, for the recognition of small molecules (e.g. O_2, NO, CO, substituted pyridines etc.) have been reported. There is, however, some debate in this grey area about what should be termed supramolecular chemistry.

Metallo-receptor for nucleic acid bases
Receptor **5.17** which extracts solid nucleobases into acetone provides an elegant example of the use of coordination chemistry to assist in binding a neutral substrate. Guanine is bound through the formation of three intermolecular interactions: a) metal ion coordination of the nitrogen atom, b) hydrogen bonding between amine protons and ether oxygen atoms and c) $\pi-\pi$ stacking (Fig. 5.21). The coordinate bond orients the substrate in a perfect manner for forming the other complementary interactions. Analogous receptors which lacked the aromatic rings did not show significant extraction of guanine, indicating the important role of $\pi-\pi$ stacking. The binding mode has been proven by crystallography. Interestingly, many enzymes use a strong coordinate interaction to support other, weaker intermolecular forces, which in turn provide binding selectivity.

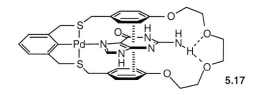

5.17

Fig. 5.21 Receptor **5.17** binds guanine using three different interactions.

Boronic acid receptors for sugars
Another recent use of dative bonds for the coordination of relatively complex neutral guest molecules has been pioneered by the research group of Seiji Shinkai in Japan. They have made extensive use of the interaction between *boronic acids* and hydroxy groups (Scheme 5.4) in order to create a range of functional receptors for sugars. Most importantly, the interaction between sugars and boronic acids readily occurs in aqueous media. High pH is required in order to facilitate the sugar complexation, but the base can also be provided intramolecularly, by the incorporation of an amine group in the receptor (e.g. **5.18**). This extends the pH range over which the receptor can operate.

The actual design of boronic acid based receptors follows a strategy with which we are, by now, familiar. Molecular clefts are synthesised in order to orient the recognition units in a *convergent* manner and, at the same time,

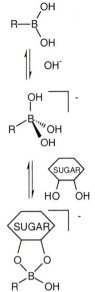

Scheme 5.3 Boronic acids interact with hydroxy groups via dative bond formation.

Table 5.4 Binding constants (log K) for receptor **5.18**.

Guest	log K
D-Glucose	3.6
D-Fructose	2.5
D-Allose	2.8
D-Galactose	2.2
1,2-ethanediol	0.2

endow the host with some *substrate selectivity*. Receptor **5.18** (Fig. 5.22) for example, shows a degree of selectivity for D-Glucose (Table 5.4). whilst other clefts prefer different saccharide guests. What is more, the receptor contains an anthracene unit, a fluorescent moiety. The fluorescent properties of this unit change on guest binding, and consequently this receptor can sense the presence of glucose molecules *via* a fluorescent response. The design of sensors, an important topic with great potential medical and environmental applications, will be discussed further in Chapter Seven.

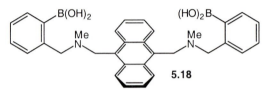

Fig. 5.22 Cleft type receptor with glucose selectivity.

The great advantage of receptors of this type, utilising relatively strong dative bonds, is that they can still operate in *competitive media*, such as aqueous solution, and therefore are of obvious potential importance and utility in biological or medical applications.

5.4 Summary and conclusions

Neutral guests provide supramolecular chemists with some of the toughest problems in the field of molecular recognition. They are often *highly functionalised*, *biologically relevant* molecules which can be addressed in a range of different ways. The role of *solvent*, which can act itself as a small neutral guest, is critical. The elegant examples discussed in this chapter, however, show the great strides which have been made into both achieving and understanding this type of molecular recognition.

Suggested further reading

For an interesting discussion of the Vancomycin story see: M.L. Gilpin, P.H. Milner, *Chemistry in Britain*, **1997**, *33*, 11, 46-48. For a general discussion of neutral molecule recognition see: A.R. van Doorn, W. Verboom, D.N. Reinhoudt in *Advances in Supramolecular Chemistry*, Ed. G.W. Gokel, Vol. 3, pp 159-206, JAI Press Inc., Greenwich, Conneticut, 1993. For an excellent overview of all 'cyclophane type' receptors see: F. Diederich, *Cyclophanes*, The Royal Society of Chemistry, Cambridge, UK, 1991. Cyclodextrin chemistry is a huge field and merits a volume of the recent encyclopaedia (useful for reference): *Comprehensive Supramolecular Chemistry*, Ed. J.L. Atwood, J.E.D. Davies, D.D. MacNicol, F. Vögtle, Vol. 3, *Cyclodextrins*, Ed. J. Szejtli, T. Osa, Elsevier Science, Oxford, 1996.

6 Self-assembly

6.1 An introduction to self-assembly

In the previous chapters we have looked at how non-covalent interactions are employed in receptors for cations, anions, ion pairs, and neutral species. In this chapter the distinction between host and guest will become blurred as we look at examples of the assembly of molecular building blocks to yield new eye-catching architectures. On mixing different appropriately designed components in solution, the intermolecular forces that exist between them control their orientation, leading to the *reversible* assembly of a specific 'supermolecule'.

There are many examples of self-assembling biological systems. The formation of the DNA double helix from two complementary deoxyribonucleic acid strands is a striking example. Under the right conditions, the thermodynamically stable double helix forms spontaneously and reversibly as the strands are mixed together and hydrogen bonds form between complementary base pairs. As a result of the reversibility of the process, any errors that may have occurred during assembly can be corrected (provided the reversibility is rapid). This type of process has been termed 'strict self-assembly.'

The formation of cell membranes, multi-component enzymes, and viruses are paradigms of the way in which Nature uses a simple, limited range of interactions to produce very complex molecular assemblies. The tobacco mosaic virus (TMV) consists of an RNA strand and 2130 protein subunits. The protein blocks self-assemble around the RNA strand *via* non-covalent interactions to form the virus superstructure (Scheme 6.1).

Self-assembly allows access to new molecular architectures that are inaccessible (or accessible in only very small yields) *via* traditional multi-step covalent bond making and bond breaking techniques. The new molecular architectures are produced by combining appropriately designed sub-units, which can be quite simple, and yet after the assembly process produce quite complex architectures. These assemblies have many potential uses ranging from information storage to drug delivery. Self-assembling systems that can self-replicate (see Section 6.4) may give us new insights into how life on earth started! It is the beauty of many of these systems, and their possible future uses (that can only currently be dreamt of in the pages of a science fiction novel), that really excite the supramolecular chemist and push this area of research forward.

This chapter is arranged so as to highlight the particular non-covalent interaction used to drive (or *template*) the formation of the assembly (such as charge transfer, metal coordination, hydrogen bonding or anion coordination).

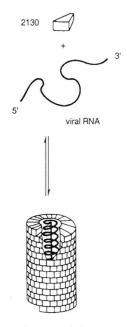

2130

viral RNA

tobacco mosaic virus

Scheme 6.1 Self-assembly of the tobacco mosaic virus (reprinted with permission from J.S. Lindsey, New J. Chem. 1991, 15, 153. Copyright (1991) CNRS-Gauthier-Villars).

When synthesizing a macrocycle there is always a chance that a precursor may thread through another macrocycle and ring close forming a catenane. This statistical approach was used to make the first synthetic catenane (Scheme 6.3). The yield of this reaction, however, is less than 1% – clearly illustrating the advantage of using self-assembly techniques in catenane synthesis!

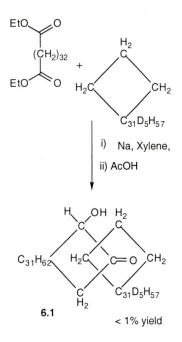

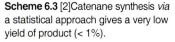

6.1

< 1% yield

Scheme 6.3 [2]Catenane synthesis *via* a statistical approach gives a very low yield of product (< 1%).

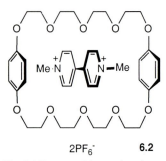

2PF$_6^-$ **6.2**

Fig. 6.1 Paraquat–bisparaphenylene-[34]crown-10 complex.

6.2 π-Electron donor–acceptor systems

Catenanes and rotaxanes

The synthesis of assemblies in which two molecular components are interlinked, but not physically joined by covalent bonds presents a considerable challenge to the supramolecular chemist. The catenanes (Scheme 6.2) and rotaxanes (Scheme 6.5) illustrate this challenge most clearly.

[*n*]Catenanes (from Latin: *catena* (chain)) consist of *n* interlocked macrocyclic species. For example a [2]catenane refers to two interpenetrating macrocycles (e.g. see Scheme 6.2). Once a catenane has formed, the only way in which the rings can be removed from one another is by breaking one of the rings open. One can regard the catenane as being held together by a *topological bond*. One of the forces used to drive the formation of such structures is the self-assembly of aromatic π-donor and π-acceptors.

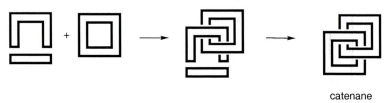

catenane

Scheme 6.2 [2]Catenane synthesis.

Stoddart and co-workers discovered that bisparaphenylene-[34]crown-10 (BPP[34]C10) will bind a paraquat dication forming a reasonably strong 1:1 complex (**6.2**) in acetonitrile and acetone solution (Fig. 6.1). This complex is stabilized by hydrogen bonding between the acidic aromatic hydrogen atoms of the paraquat and the oxygen atoms of the crown ether and also by a *charge transfer π–π stacking interaction* from the *electron rich crown ether* to the *electron poor paraquat cation*. Similarly electron poor macrocycles have been synthesized which will encapsulate electron rich guests. This type of interaction has been used to synthesize a wide range of rotaxanes and catenanes. The synthesis of a [2]catenane from BPP[34]C10 and compound **6.3** is shown in Scheme 6.4. The pyridine nitrogen of compound **6.3** acts as a nucleophile, displacing a bromine from 1,4-bis(bromomethyl) benzene. The resultant tricationic strand (**6.4**) assembles with the crown ether present in the reaction mixture, threading through it to form a complex that is stabilized by hydrogen bonding and π–π stacking. This complex is then trapped by a ring closing intramolecular nucleophilic attack to form a [2]catenane **6.5** (isolated as the hexafluorophosphate salt). The crown ether may be regarded as a template around which the pyridinium box forms. This template effect is reflected in the high yield (70%) of the reaction.

Catenanes with larger numbers of rings have been produced by analogous methods. The [5]catenane **6.6** was dubbed *olympiadane* due to its similarity to the five interlocked olympic rings. It provides an elegant example of the power of self-assembly and the use of non-covalent interactions to access complex molecular arrays with high molecular mass (Fig. 6.2).

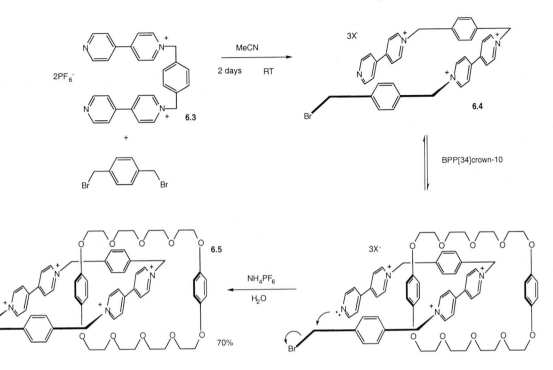

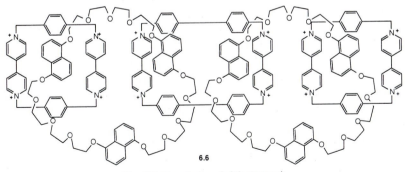

Scheme 6.4 [2]catenane formation.

Rotaxanes (from Latin: *rota* (wheel) and *axis* (axle)) are composed of a macrocyclic component through which an axle or rod is threaded. The ends of the axle are stoppered with bulky groups that prevent the macrocycle slipping off. Therefore the individual components are permanently joined together, but are not linked by an actual covalent bond. The nomenclature [*n*]rotaxane refers to the total number of non-covalently linked components present in the rotaxane. Thus a [2]rotaxane consists of one macrocycle and one axle.

Fig. 6.2 Olympiadane (a [5]catenane).

Strategies for rotaxane synthesis are illustrated in Scheme 6.5. There are two main approaches: *threading* and *clipping*. Threading involves mixing self-assembling linear and macrocyclic components to form a pseudorotaxane. The ends of the linear component are then stoppered to prevent the wheel slipping off the axle, producing a [2]rotaxane. For clipping, a 'pre-stoppered' linear component may be mixed with a self-assembling component which undergoes macrocyclic ring closure around the axle. Scheme 6.6 shows the synthesis of rotaxane **6.10** *via* both of these routes.

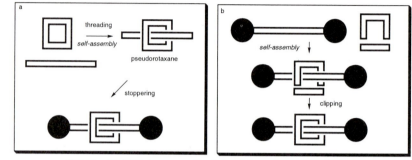

Scheme 6.5 [2] Rotaxane synthesis by a) threading and b) clipping.

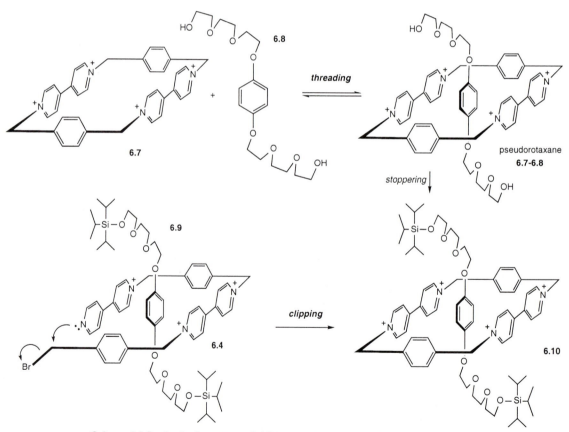

Scheme 6.6 Synthesis of a rotaxane (**6.10**) by two different routes: threading and clipping.

A third approach to rotaxane synthesis has also been developed by Stoddart's research group. They reasoned that if the stoppers are not too large, then at high temperatures, the macrocyclic component may be able to slip over them. The rotaxane is thermodynamically stable compared to the unassembled component parts due to the favourable π-donor–π-acceptor interaction and hydrogen bonding interactions (Fig 6.3). Enough thermal energy must be input in order to traverse the activation barrier to threading.

More complicated rotaxane systems have been produced, for example [3]rotaxanes (one axle with two macrocycles). Axles containing two possible positions (stations) for one macrocyclic component have been produced and can be considered as molecular shuttles, with the position of the macrocycle on the rod being variable and controllable (Scheme 6.7). Readers are urged to consult the further reading section at the end of this chapter for more information on these fascinating molecules.

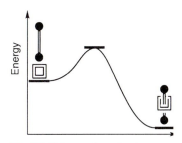

Fig. 6.3 Rotaxane formation by slippage requires elevated temperatures.

6.3 Transition metal directed assemblies

The use of transition metal ions to direct molecular assembly has two major advantages. Firstly, metal-ligand dative bonds are *thermodynamically strong* interactions, but have *varying degrees of lability* (providing the supramolecular chemist with a range of kinetic stabilities). They can therefore provide stabilisation energy for a range of different structures. Secondly, due to ligand field effects, transition metal ions often have very *specific geometric requirements* in their coordination sphere. This gives metal ions the ability to control the geometry of the molecular assembly very precisely. This process can be regarded as an extension of the template effect discussed in Chapter 2.

Catenates and catenands

Metal directed assembly also provides an approach to interlocked structures complementary to that described in Section 6.2. Sauvage and his research group have investigated the assembly properties of 2,9-disubstituted 1,10-phenanthroline molecules with copper(I) ions.

Scheme 6.8 shows the synthesis of a [2]catenate by a metal directed route starting from a 2,9 phenol-substituted 1,10-phenanthroline unit. Copper(I) has a d^{10} electronic configuration and prefers a tetrahedral coordination environment. When phenanthroline **6.11** is mixed with 0.5 equivalents of copper (I), the two phenanthroline units are locked together forming the tetrahedral complex **6.12**. The subsequent nucleophilic substitution reaction with pentaethylene glycol diiodide forms the interlocked catenate **6.13** in 27% yield. This structure is called a catenate rather than a catenane because it contains a metal ion. Demetallation of the catenate is possible by addition of CN⁻ yielding the [2]catenand **6.14**.

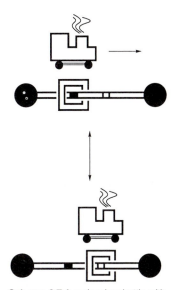

Scheme 6.7 A molecular shuttle with two stations.

A *catenand* is a catenane that is capable of coordinating to a metal ion. The metal complex is referred to as a *catenate*.

Double and triple helices

The formation of double and triple helices presents many challenges to the supramolecular chemist. The helices described in this section consist of ligands containing multiple binding sites that self-assemble around metal ions. The formation of helices is permitted in assemblies containing more than one metal ion (Scheme 6.9). As will be illustrated, both the arrangement of the binding sites on the ligand and the preferred coordination geometry of the metal ion are crucial to successful helicate synthesis.

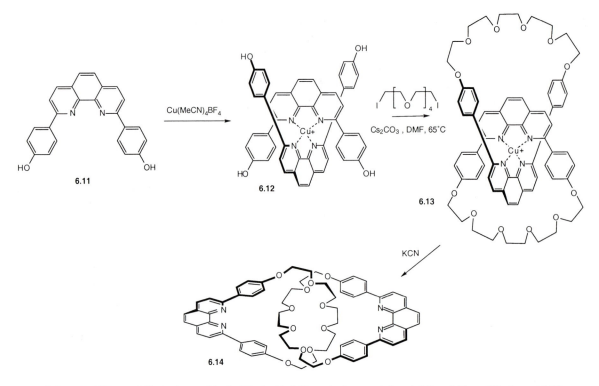

Scheme 6.8 The metal directed assembly of a [2]catenate **6.13** and subsequent demetallation yielding a [2]catenand **6.14**.

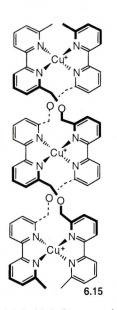

Fig. 6.4 A double helicate containing three copper (I) metal ions.

Jean-Marie Lehn's group at Strasbourg produced a series of polymetallic double helices based on poly-bipyridine ligand strands with ether linkages between each bipyridine subunit. An example containing three copper (I) ions is shown in Fig. 6.4. Once again, the copper (I) ions enforce a tetrahedral coordination geometry that causes the assembly of the helical structure. Helicates with up to five copper ions were produced

Lehn found that if poly-bipyridine strands containing two, three, four, and five bipyridine units were mixed together then, in the presence of copper (I) ions, *only helices containing two strands of the same length would form.* Lehn termed this remarkable process *'self-recognition.'*

Metal ions with octahedral coordination environments may also be found in helicates as illustrated in Scheme 6.9. Two strands of the quaterpyridine ligand **6.16** wrap themselves around two copper(I) ions forming double helical **6.17**. In this way, all of the nitrogen donor sites on the ligand strands are involved in binding metal ions and the ions are adopting their preferred coordination geometries. The quinquepyridine strand **6.18**, however, possesses five donor nitrogen atoms. Clearly, a double helix with copper(I) ions analogous to **6.17** cannot form, as there would be two unsatisfied nitrogen donor atoms. If, however, this ligand is mixed with copper(I) ions and then subjected to mild oxidizing conditions, a different double helical structure forms, in this case a mixed valence Cu(I)-Cu(II) helicate **6.19**. Copper(II) is d^9 and therefore prefers distorted octahedral coordination environments (due to the Jahn-Teller effect). In **6.19**, therefore, the copper

(II) ion is coordinated to three nitrogens from each quinquepyridine strand and so is in a favourable 6-coordinate environment. The Cu(I) ion is in a favourable tetrahedral coordination site occupying the remaining two nitrogens on each quinquepyridine. In this way, both the metal ions are in their preferred geometries and all the donor atoms of the ligand are satisfied. The helicates **6.17** and **6.19** may twist in either a left or right handed manner and are therefore both chiral assemblies.

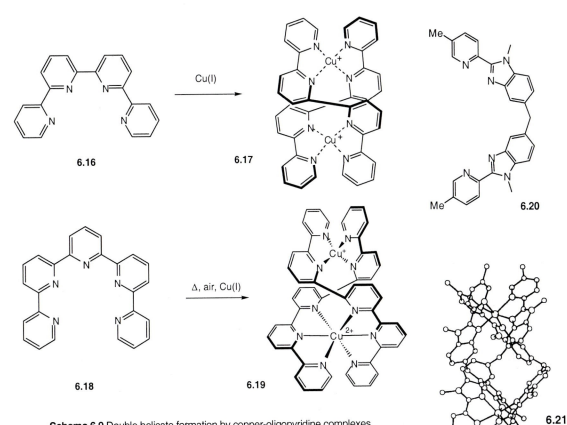

Scheme 6.9 Double helicate formation by copper-oligopyridine complexes.

Benzimidazole-pyridine ligands (e.g. **6.20**) form different types of helicate depending upon the metal ion used to template the self-assembly process. The arrangement of nitrogen atoms in **6.20** provides two metal binding sites, while the inflexibility of the strand forces the formation of dimetallic complexes. The addition of copper(I) ions causes the assembly of a double helical structure, with four coordinate metal ions. However, addition of cobalt (II) ions causes the assembly of a *triple helicate* **6.21** (because only then can the metal ions achieve their preferred six-coordinate octahedral geometries (Fig. 6.5)). This illustrates how metal ion coordination geometry requirements can control the structure of the assembly.

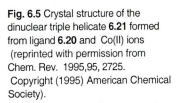

Fig. 6.5 Crystal structure of the dinuclear triple helicate **6.21** formed from ligand **6.20** and Co(II) ions (reprinted with permission from Chem. Rev. 1995, 95, 2725. Copyright (1995) American Chemical Society).

Knots

The trefoil knot was an ambitious chemical goal (Scheme 6.10). Such a knot contains one single strand which intertwines itself with three crossing points, such that it can not be unwound. Sauvage and his research team reasoned that such a molecule could be approached by combining the concepts of catenane synthesis with the formation of bi-metallic helical complexes (Scheme 6.10). Initially a double helical structure is assembled around two metal ions. The ends of the helix are then linked together in a cyclization reaction similar to that used in catenate formation and subsequently the metal ions may be removed. In contrast to the catenate topology that is generated when one metal centre is used (Scheme 6.8) the cyclization step in Scheme 6.10 produces not two interlocked rings but one continuous ring that crosses over itself in three places: a trefoil knot.

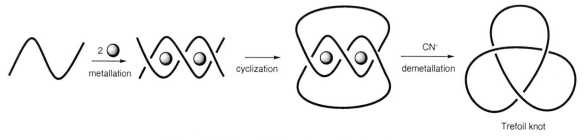

Trefoil knot

Scheme 6.10 The metal directed assembly of a trefoil knot.

The synthesis of a trefoil knot is shown in Scheme 6.11. The metal ligating strand **6.22** contains two phenanthroline units linked by a 1,3-phenylene spacer group. In the presence of copper(I) ions a double helical complex **6.23** forms which is subsequently cyclized by reaction with pentaethylene glycol diiodide in the presence of Cs_2CO_3 affording the metallated trefoil knot **6.24** in 29% yield. The copper (I) ions may then be removed by addition of KCN giving trefoil knot **6.25**.

Molecular macrocycles and boxes

Research into the self-assembly of macrocycles and boxes via metal ion coordination has produced some beautiful molecular architectures (Schemes 6.12 and 6.13) which have the potential to bind other species within their central cavities.

The porphyrin-pyridine conjugate **6.26** is self-complementary due to the subtle use of hydrogen bonds. Pyridine-amide hydrogen bonds (shown as dotted lines in Scheme 6.12) force the porphyrin and terminal pyridine to be orientated at 90° to one another. This allows the formation of the rectangular macrocyclic species **6.27** *via* pyridine coordination to the zinc metal ion bound in the porphyrin ring.

Jeremy Sanders' group at Cambridge University have pioneered the use of self-assembly in the synthesis of boxes containing porphyrin binding sites. Addition of the tri-coordinating 1,3,5-tris(4-pyridyl)triazine **6.29** to the dialkyne metalloporphyrin **6.28** and subsequent copper catalysed coupling of the alkyne groups forms the trimeric macrocycle **6.30** (Scheme 6.13). By replacing **6.29** with di-coordinating 4,4'bipyridine, a dimeric macrocycle is predominantly formed, indicating that the added pyridine molecule is coordinating to the metal ions and templating the formation of the macrocycle (therefore strictly speaking this is a ligand directed assembly). The trimer has been used as a molecular reaction vessel with Diels-Alder reactions occurring (and being catalysed) within the central cavity (see Section 7.5).

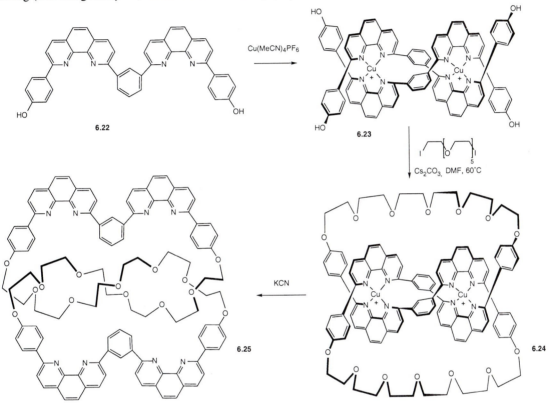

Scheme 6.11 Sauvage's trefoil knot synthesis.

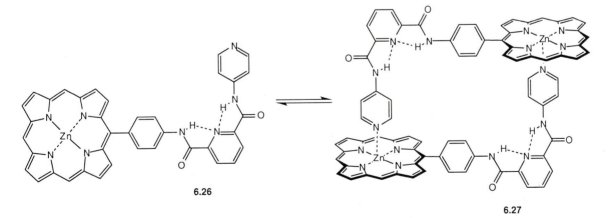

Scheme 6.12 Assembly of a porphyrin based molecular box (substituents on porphyrin rings are not shown).

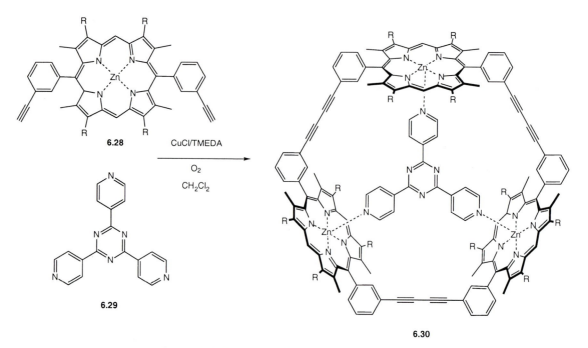

Scheme 6.13 Metal assisted synthesis of a triporphyrin box.

Locked and unlocked molecular boxes

The kinetic lability of a metal ion can determine whether a molecular box is 'locked' or 'unlocked'. The Pd(II) cation prefers square planar coordination and has been employed in the assembly of a number of different molecular boxes which possess 90° angles at the corners (for example see Scheme 6.14). Mixing ligand **6.31** with Pd(en)(ONO$_2$)$_2$ produces a mixture of catenate **6.32** and macrocycle. **6.33**. The nitrogen-palladium bonds are labile and so an equilibrium exists between the macrocycle and catenate (i.e. the molecular box is unlocked). At low concentrations the macrocycle predominates whilst at high concentration the catenate is the dominant component of the equilibrium.

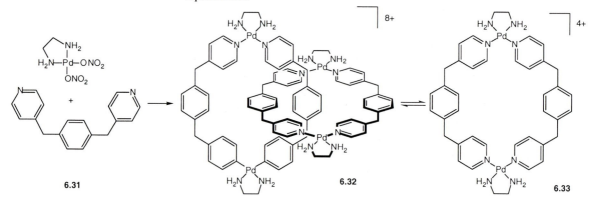

Scheme 6.14 Palladium box-catenate equilibrium.

Analogous systems containing Pt(II) e.g. **6.34** are not labile under ordinary conditions and may be isolated as discrete entities (locked boxes). However, under very polar conditions and on heating, the Pt-N bonds become labile and this may be used to assemble catenates that are kinetically stable. When **6.34** is heated at 100°C with 5M NaOH the bond becomes labile (is unlocked), allowing formation of the [2]catenate (**6.35**). When the reaction mixture is cooled and the salt is removed the Pt-N bonds become locked allowing isolation of the kinetically stable [2]catenate (Scheme 6.15).

Racks, ladders and grids

Metal directed assembly allows the formation of increasingly complex molecular superstructures. A number of these structures are shown schematically in Fig. 6.6.

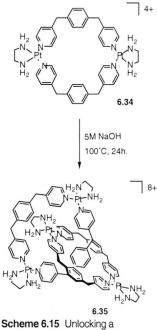

Scheme 6.15 Unlocking a macrocycle to form a catenate.

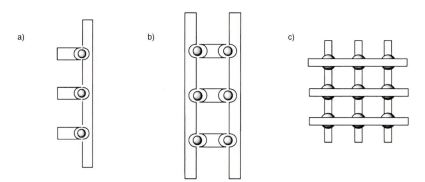

Figure 6.6 A schematic representation of a molecular rack (a), a ladder (b) and a grid (c).

A *rack* structure (Scheme 6.16a) is formed when the tris-bipyridine strand **6.36** is mixed with the 1,10-phenanthroline crown ether **6.37** in the presence of copper (I) ions. This rack-like complex **6.38** is actually a pseudo-rotaxane with three macrocycles bound to the central axle *via* three copper (I) metal ions.

However when a bipyridine strand is mixed with six copper(I) ions and three bispyrimide units **6.39** a *ladder* forms (**6.40**) (Scheme 6.16b).

Spectacular complexes such as the 3 x 3 *molecular grid* **6.42** may be produced by mixing the linear ligand **6.41** with nine silver(I) ions (Scheme 6.16c). Lehn has suggested that arrays of metals, such as that present in **6.42**, may be used in the future for information storage. One could imagine each metal ion as corresponding to a 'bit' of information (for example one oxidation state could correspond to 'on' and another to 'off'). Such arrays would allow the storage of large amounts of information in very small volumes of material, however we need to find a way of *addressing* (reading and writing information from and to) the individual metal ions first! This type of approach to functional materials is often termed a *nanoscale approach*.

The examples discussed above illustrate the versatility and controllability of metal coordination interactions in the construction of unique architectures.

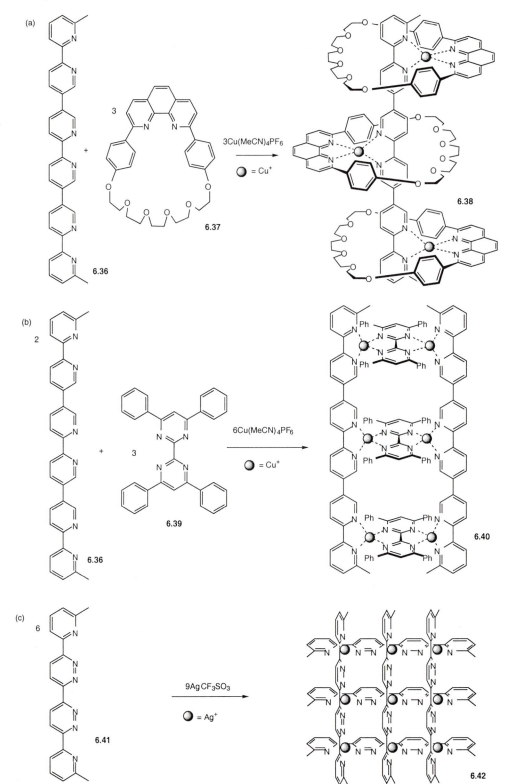

Scheme 6.16 Metal directed assembly of a) a rack, b) a ladder, and c) a grid.

6.4 Hydrogen bond directed assemblies

The *directional nature* of hydrogen bonds makes them well suited for the assembly of complex supermolecules. Indeed hydrogen bonding is key to the assembly of perhaps the most beautiful natural supermolecule, the DNA double helix. Supramolecular chemists have also employed hydrogen bonding motifs in the construction of a wide variety of assemblies, a selection of which are described below.

Rosettes and ribbons

Hydrogen bonds have been used to produce supermolecules incorporating melamine and cyanuric acid derivatives. Melamine **6.43** can be regarded as having three faces, each with a donor-acceptor-donor (DAD) triad of hydrogen bonding groups. This is a complementary arrangement to that of cyanuric acid **6.44** which possesses an ADA triad. When these two species are mixed together in solution, an insoluble polymeric complex with a 'rosette-like' arrangement of sub-units (**6.45**) precipitates (Scheme 6.17). This complex is held together by the multiple hydrogen bonds between adjacent heterocycles.

By attaching three derivatized melamine units to a trigonal template (a 1,3,5-substituted benzene ring), a soluble 3:1 complex **6.48** was formed with cyanuric acid (Scheme 6.18). The existence of this assembly was ultimately proven by NOESY nmr experiments, whilst vapour phase osmometry was used to provide an approximate molecular mass for the 3:1 stoichiometry.

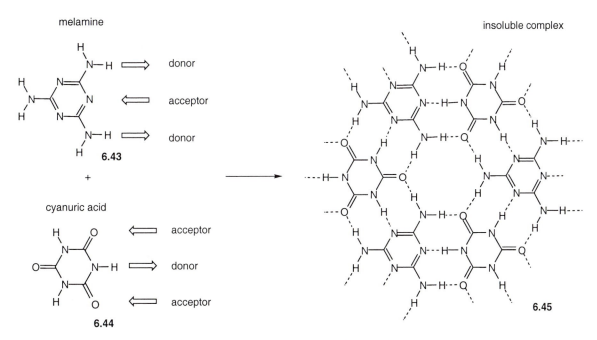

Scheme 6.17 Formation of a cyanuric acid - melamine complex.

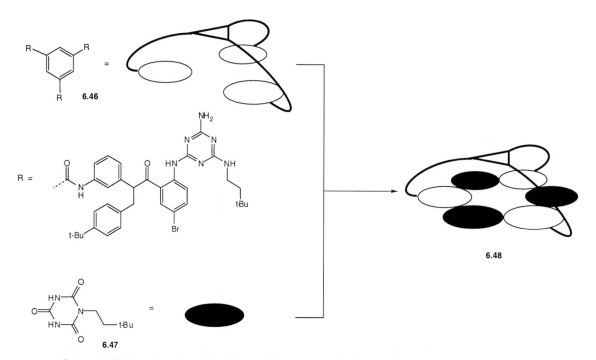

6.46

6.48

6.47

Scheme 6.18 Formation of a soluble 3:1 complex between a melamine tripod and a cyanuric acid derivative.

A variation on this hydrogen bonding motif was used by Lehn and co-workers to synthesize molecular ribbons. Compound **6.49** is a barbituric acid derivative, that is similar to compound **6.44** except that one of the hydrogen bonding faces of the molecule has been blocked by two alkyl groups. Compound **6.50** is similar to melamine except that an octyl group prevents hydrogen bonding on one face of the heterocycle. Co-crystallization of **6.49** and **6.50** produces a molecular ribbon **6.51** with the bulky blocking groups protruding from each side of the linear array (Scheme 6.19). This type of ribbon can be considered as a *supramolecular polymer*, with the length of the aggregate being dependent on the strength of the hydrogen bonding interactions. Stronger association between the individual components would lead to the formation of longer ribbons.

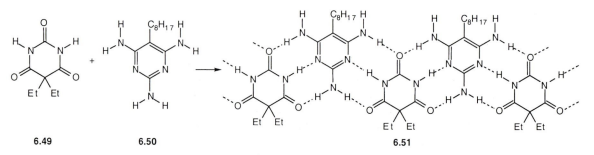

6.49   **6.50**   **6.51**

Scheme 6.19 Formation of a molecular ribbon (supramolecular polymer).

Hydrogen bonded rotaxanes and catenanes

Hydrogen bonds have also been used to drive the formation of rotaxanes and catenanes.

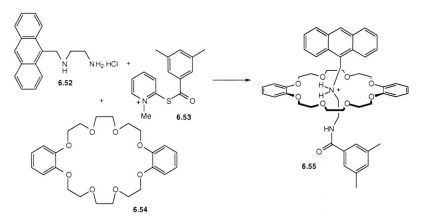

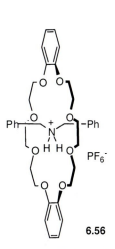

Scheme 6.20 Hydrogen bond directed assembly of a rotaxane.

One example is shown in Scheme 6.20. Acylation of the anthracene-amine **6.52** with **6.53** in the presence of dibenzo[24]crown-8 affords the rotaxane **6.55** in 22% yield in a mixture of chloroform and water. The rotaxane is stabilized by hydrogen bonding and electrostatic interactions between the ammonium group and the crown ether and also by π–π stacking interactions between the anthracene stopper and a benzo group in the crown ether. The same strategy has been employed in assembling pseudorotaxanes. Complex **6.56** forms spontaneously when dibenzylammonium ions are mixed with dibenzo[24]crown-8 (Fig. 6.7).

Catenanes may also be formed *via* hydrogen bond directed assembly. By altering the synthetic route used to prepare his benzoquinone receptor (see Section 5.1), Hunter has prepared the [2]catenane **6.58** in 34% yield (Scheme 6.21) by reaction of **6.57** with isophthaloyl chloride. Hydrogen bonds hold two molecules of **6.57** together perpendicular to one another in a semi-interlocked manner, and then isophthaloyl chloride completes macrocycle formation leaving the rings permanently interlocked. The catenane is stabilized by a combination of π–π stacking and hydrogen bonding interactions.

Peptide nanotubes

Ghadiri and his co-workers at The Scripps Research Institute in California have recently reported the self-assembly of cyclic peptides *via* multiple hydrogen bonds. The peptides consist of a cyclic array of alternating L- and D-amino acids that adopt a conformation such that amide hydrogen bond donor N-H and acceptor C=O groups are oriented perpendicularly to the plane of the macrocycle. Each cyclic peptide can form twenty four hydrogen bonds (twelve 'up' and twelve 'down'), a process which enthalpically drives formation of the tube architecture (**6.59**) (Fig. 6.8). The assembly of these nanotubes is tunable, with different sizes of cyclic peptide leading to nanotubes of different

6.56

Fig. 6.7 A pseudo[2]rotaxane formed *via* favourable hydrogen bonding interactions.

diameter. These nanotubes are especially exciting because of the possibility of transporting molecules through the channel. Nanotubes constructed from hydrophobic amino acids have been suspended in lipid bilayer membranes and shown to transfer glucose molecules through the membrane.

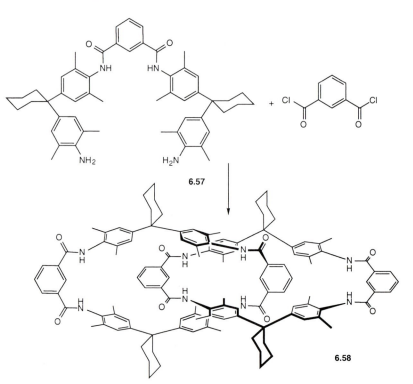

6.57

6.58

Scheme 6.21 Hunter's amido-[2]catenane.

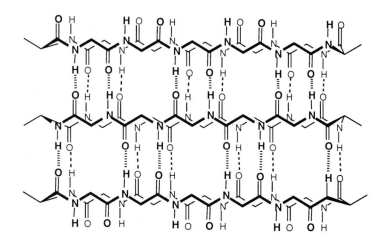

Figure 6.8 Ghadiri's peptide nanotube **6.59**.

The tennis ball: encapsulation in a self-assembled capsule

Rebek and co-workers have synthesized the self-complementary unit **6.60** by condensing two molecules of glycoluril with durene tetrabromide. This building block is curved due to steric hindrance between the large number of phenyl groups protruding from one face of the molecule. Two molecules of **6.60** self-assemble in solution *via* the formation of eight hydrogen bonds to make a 'tennis-ball' like molecular capsule **6.61** (Scheme 6.22). The assembled tennis ball is hollow allowing encapsulation of small molecules. Methane and xenon are two guests that have been captured by the use of the van der Waals interactions.

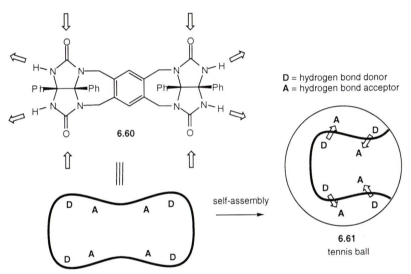

D = hydrogen bond donor
A = hydrogen bond acceptor

6.60

self-assembly

6.61
tennis ball

Scheme 6.22 The formation of a molecular tennis ball

Self-replicating molecular systems

The design of molecules that are capable of replication is a research area of great interest to chemists studying the role of self-assembly in the evolution of early life. The concept is to create molecules that upon synthesis, actually enhance their own formation. In other words, the synthesis can feed back and the product molecule can be thought of as making additional copies of itself (self-replication). One such system is shown in Scheme 6.23. In this case the formation of an imine bond is catalysed by an orientational effect caused by the formation of amidinium-carboxylate salt bridges.

Reaction of 2-formylphenoxyacetate **6.62** with 3-aminobenzamidinium **6.63** produces the product **6.64**. **6.64** contains both carboxylate and amidinium groups and so forms a complex with the two individual starting materials **6.62–6.64–6.63**. In this complex the amine and aldehyde groups on reagents **6.62** and **6.63** are held close together, increasing the likelihood of nucleophilic attack leading to imine bond formation (the reaction effectively becomes intra-molecular — reducing the entropy costs of bond formation). This system is self-replicating because the product acts as a catalyst for its own formation (autocatalysis).

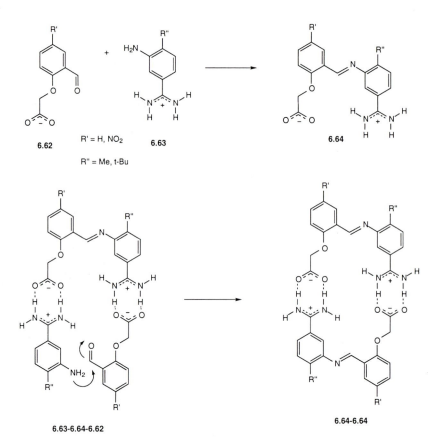

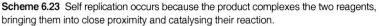

Scheme 6.23 Self replication occurs because the product complexes the two reagents, bringing them into close proximity and catalysing their reaction.

6.5 Anion directed assemblies

The role of anions in promoting the formation of self-assembled arrays has only recently begun to be studied. As we have seen in Chapter 3, anions are more challenging to recognize than cations because of their more diffuse nature (small charge to radius ratio), pH sensitivity and generally wider range of geometries. This might explain why very few examples of anion directed assemblies have been reported.

A striking example of anion directed assembly is shown in Scheme 6.24. This pentametallic circular helicate **6.65** forms when the tris-bipyridine ligand shown in the complex is mixed with an equimolar amount of $FeCl_2$ in ethylene glycol at 170°C. The structure is related to the helices shown earlier in the Chapter, with interwoven bipyridine strands coordinated to octahedral metal ions. The chloride anion bound in the centre of the helicate cannot be exchanged for other anions such as PF_6^- or $CF_3SO_3^-$ demonstrating the selectivity of the helicate for Cl^-. If another iron salt is used in the reaction, such as $Fe(BF_4)_2$, the pentametallic structure *does not form* and instead a hexameric complex is obtained. This clearly demonstrates the role played by the chloride anion in templating the assembly of the pentamer.

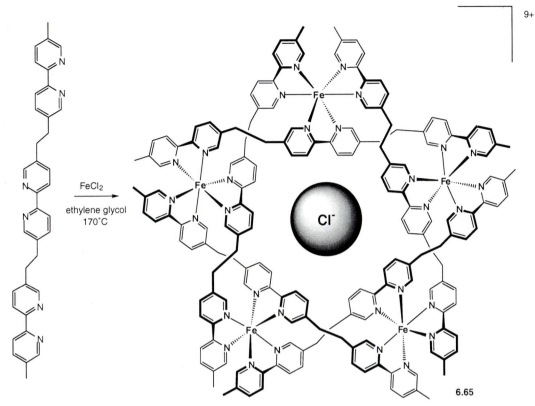

Scheme 6.24 A circular double helicate **6.65** formed around a chloride anion.

An example of anion directed assembly of a nickel cage complex has been reported by Mingos and co-workers at Imperial College, London. Reaction of $NiCl_2$ with amidinothiourea **6.66** (Fig. 6.9) in methanol yields crystals of the cage complex **6.67** shown in Fig. 6.10.

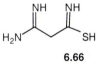

Fig. 6.9 Amidinothiourea.

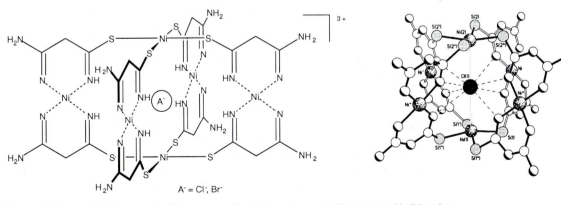

Figure 6.10 An amidinothiourea-nickel(II) cage complex **6.67** spontaneously forms around halide anions. The crystal structure (right) shows a chloride anion bound by eight NH—Cl hydrogen bonds (reprinted with permission from Angew. Chem., Int. Ed. Engl. 1998, 37, 1258. Copyright (1998) WILEY-VCH).

The cage $[Ni_6(atu)_8Cl]Cl_3$ consists of eight amidinothiourea units that coordinate six nickel ions through both nitrogen and sulfur donor atoms. A single chloride anion is bound in the centre of the cage *via* eight NH–Cl hydrogen bonds. An analogous cage complex can also be formed using $NiBr_2$. However when nitrate, acetate or perchlorate salts are used, the simple monomer $[Ni(atu)_2]^{2+}$ complexes result. When chloride anions are subsequently added to these complexes (in the form of KCl) the cage complex spontaneously forms around the halide.

6.6 Summary and conclusions

The principles of supramolecular chemistry can be applied to the crystalline state. Hydrogen bonding, metal-ligand coordination and charge transfer interactions are currently being used in an attempt to predict and control the formation of novel three dimensional lattice frameworks. By appropriate design, these zeolite-like materials may exhibit novel inclusion, catalytic, magnetic or electronic properties. This area of chemistry is known as 'crystal engineering' and is unfortunately beyond the scope of this book. Interested readers are however urged to consult the further reading section at the end of this chapter.

The use of supramolecular methodology in order to control intermolecular interactions has led to a stunning array of different molecular architectures: catenanes, rotaxanes, helices, knots, macrocycles, racks, ladders, grids, tubes... the structures synthesized seem limited only by the imagination of the chemist. The 'building block' approach to these assemblies allows the synthesis of structures which would be inaccessible or non-viable *via* traditional organic synthesis, which focusses on the formation of individual covalent bonds. What is more, many of the self-assembled systems discussed in this chapter are now exhibiting interesting functions ranging from guest encapsulation or transport through to storage of information at the nanometre level and the self-assembly of zeolite type materials (see margin). The future for self-assembled supramolecular systems appears very bright indeed.

Suggested further reading

For examples of self-assembly in biological systems, see: D. Voet, J. G. Voet, Biochemistry, 2nd ed., Wiley, New York, 1995, C. Branden, J. Tooze, Introduction to Protein Structure, Garland, New York, 1991 and A. Fersht, Enzyme Structure and Mechanism, 2nd ed., Freeman, New York, 1985. For some definitions of self-assembly see: J. S. Lindsey, *New Journal of Chemistry* **1991**, *15*, 153-180. For an excellent collection of reviews on self-assembling systems, see: Comprehensive Supramolecular Chemistry, vol. 9 (Templating, Self-assembly, and Self-organization), Ed. J. L. Atwood, J. E. D. Davies, D. D. MacNicol, F. Vögtle, Elsevier, Oxford, 1996. For two outstanding review articles on self-assembling systems see: "Interlocked and Intertwined Structures and Superstructures" D.B. Amabilino and J.F. Stoddart, *Chemical Reviews* **1995**, 95, 2725-2828 and "Self-assembly in Natural and Unnatural Systems" D. Philp and J.F. Stoddart, *Angewandte Chemie, International Edition in English* **1996**, 35, 1154-1196. For information on crystal engineering see: Comprehensive Supramolecular Chemistry, vol. 6 (Solid-state Supramolecular Chemistry: Crystal Engineering), Ed. J. L. Atwood, J. E. D. Davies, D. D. MacNicol, F. Vögtle, Elsevier, Oxford, 1996.

7 Present & future applications

This final chapter examines some of the applications of supramolecular chemistry as well as speculating on what the future may hold for molecular assemblies. This includes new separation media for use in chromatography, novel sensors, new catalysts, new pharmaceuticals and even the creation of nanoscale molecular machinery.

7.1 Phase transfer agents

In Chapter 2, we saw that crown ethers can *solubilize* salts in non-polar solvents. This has allowed their use as phase transfer agents in a number of organic reactions. For example, KF may be solubilized in acetonitrile by the addition of [18]crown-6. In this case, once in solution, the fluoride anion is relatively unsolvated and therefore strongly nucleophilic. (Scheme 7.1). Similarly, potassium permanganate is solubilized in benzene by dicyclohexyl-[18]crown-6. So called 'purple benzene' is a powerful oxidizing reagent as a result of benzene's inability to solvate and stabilize the anion (Scheme 7.2).

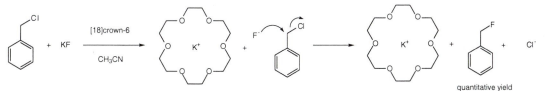

quantitative yield

Scheme 7.1 Crown ethers act as phase transfer catalysts for nucleophilic substitutions.

Applications of alkali metal cation phase transfer also bring us back to one of the first examples in this book: valinomycin. This antibiotic operates by selectively transporting potassium cations through the cell membrane phase.

7.2 Separation of mixtures

The production of new materials that allow the separation of one species from a mixture containing many components is of prime importance. For example, removal of pollutants such as toxic metal ions from aqueous solutions is environmentally important. This process involves molecular recognition and therefore lies at the heart of supramolecular chemistry.

Solid supports

By attaching a receptor to a solid support we create a modified material that can be used for the *removal* of a particular substrate from a mixture *e.g.* by adding the modified solid support to a solution containing the target substrate and then filtering off the solid support with the substrate bound to it.

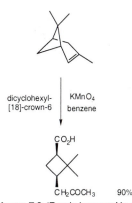

dicyclohexyl-[18]-crown-6 | KMnO$_4$ benzene

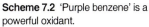

CH$_2$COCH$_3$ 90%

Scheme 7.2 'Purple benzene' is a powerful oxidant.

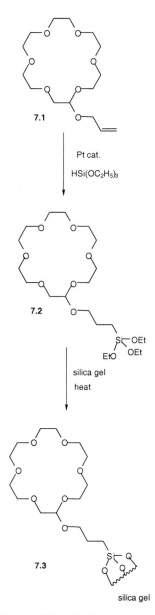

7.1

Pt cat.

HSi(OC₂H₅)₃

7.2

Si—OEt

EtO   OEt

silica gel

heat

7.3

silica gel

Scheme 7.3 Synthesis of an [18]crown-6 modified silica gel.

Alternatively the modified solid support may be used as a *separation medium* in chromatography. The modified support is packed into a column and mixtures of substrates passed over it. In this way, separations may be achieved according to the differences in the binding constants of the various substrates with the solid bound receptor species.

Izatt, Bradshaw and Christensen have produced a number of such materials designed to selectively remove or concentrate metal ions in solution, for example silica gel modified with [18]crown-6 (Scheme 7.3). The metal ion selectivity of the silica bound crown ether was found to be similar to the free crown ether. Aqueous solutions containing mixtures of cations were passed through this material in order determine the selectivity of the gel for particular cations. It was found that Ba^{2+} was bound 10 times more strongly than Sr^{2+} and 339 times more strongly than Ca^{2+}. Therefore when a mixture when a mixture of these cations is run through the crown ether modified silica gel, Ca^{2+} is bound most weakly and emerges from the column first. Ba^{2+} is bound most strongly and emerges from the column last.

Using the same strategy, these researchers have attached thiacrown ethers to silica gels and produced materials that are capable of removing precious and/or toxic metals from aqueous solutions. Binding constants for competitive, undesired base metals such as Cu^{2+} and Fe^{3+} were at least 10 orders of magnitude less than constants for precious Au^{3+} and Ag^+, or toxic Pd^{2+} and Hg^{2+}. An area of intense current activity is the development of solid supported extraction agents for the separation of radioactive metals. Such extraction media would allow removal of uranium and plutonium residues from power station waste, enabling a more efficient treatment of nuclear waste.

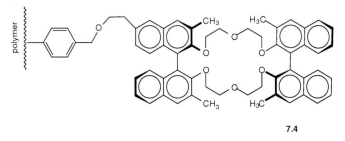

7.4

Fig. 7.1 a) (*RR*)-binaphthyl crown ether bound to a polystyrene resin

Cram and co-workers have attached chiral crown ethers to polystyrene resin and used the resulting materials to achieve chiral chromatographic separations of α-amino acids and ester salts. An example of an *RR*-binaphthyl crown ether modified support is shown in Fig. 7.1. Chiral substrates passing through a column made of this material interact with the crown ethers forming diastereomeric complexes which have different stability constants. The *R*-enantiomer forms a thermodynamically more stable complex with the host and is more strongly bound to the column. Therefore, the *S*-enantiomer emerges from the column first and resolution of the enantiomers is achieved (Fig. 7.2). Chiral chromatographic separations are of great interest to the pharmaceutical industry, which often has to resolve two enantiomers of a chiral drug which possess completely different activities.

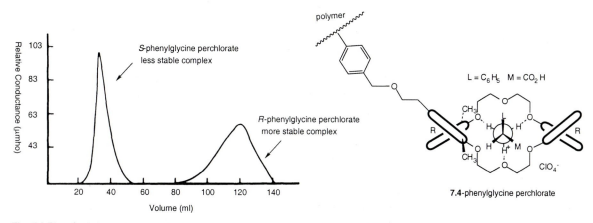

Fig. 7.2 Plot of relative solution conductance vs. volume of column eluent for phenylglycine perchlorate on Cram's *(RR)*-binaphthyl column eluting with chloroform/acetonitrile (left) and a schematic of the phenylglycine perchlorate complex (right). (Reprinted with permission from J. Am. Chem. Soc., 1976, 98, 3038, Copyright (1976) American Chemical Society).

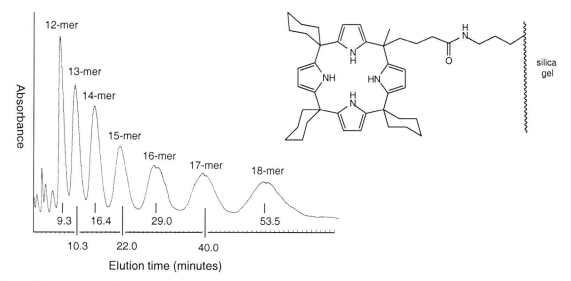

Fig. 7.3 The separation of anionic oligonucleotides $(dT)_{12-18}$ on calixpyrrole modified silica (reprinted with permission from Chem. Commun. 1998, 1, Copyright 1998 The Royal Society of Chemistry).

This technique is not limited to separating mixtures of cations. Very recently, calixpyrrole (see Chapter 3) modified silica gels have been used to separate anionic mixtures. Figure 7.3 shows the separation of short nucleic acid strands (oligonucleotides) with such a gel. The oligonucleotides range from 12 to 18 bases in length and are anionic (they contain negatively charged phosphate groups). The longer oligonucleotides (e.g. the 18-mer) are retained on the column longer than the shorter oligomers. This reflects the fact that the longer oligonucleotides have more phosphate groups and therefore a higher chance of interacting with a particular calixpyrrole.

Calixarene purification of C$_{60}$

The production of pure Buckminsterfullerene C$_{60}$ in reasonable quantity required lengthy column chromatographic procedures before Atwood and Shinkai independently discovered that *p-tert*-butylcalix[8]arene could be used to separate the spherical C$_{60}$ soccer ball from rugby ball C$_{70}$ shaped impurities (Scheme 7.4). C$_{60}$ selectively forms a complex with *p-tert*-butylcalix[8]arene in toluene. The spherical C$_{60}$ molecule sits in the cup shaped cavity of the calix[8]arene and the complex precipitates from solution. C$_{70}$ and other impurities do not form complexes with the calixarene and so remain in solution. The C$_{60}$ complex is isolated by filtration. When this complex is suspended in chloroform it dissociates: the calixarene dissolves whereas the C$_{60}$ remains insoluble and may be isolated by filtration.

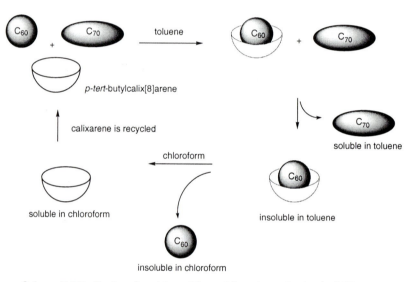

Scheme 7.4 Purification of a mixture of C$_{60}$ and C$_{70}$ using *p-tert*-butylcalix[8]arene.

7.3 Molecular sensors

A receptor may be used as a sensor if it can report the presence of the guest by some physical means. A sensor should ideally be selective for a particular guest and not only report the presence of the guest molecule, but should also allow the chemist to monitor its concentration. This is important medically (for monitoring indicators of physiological function) and environmentally (monitoring pollutant levels). Two different strategies have been applied to sensor production.

Firstly, the receptor can be used to create a modified material, for example an electrode. The receptor is incorporated into a polymer electrode, and this modified electrode can then show a selective response to the presence of the ion for which the receptor is selective, allowing the quantitative determination of ion concentrations in solution. This important topic is included in another primer text (Chemical Sensors by R.W. Cattrall) to which the interested reader is directed for further information.

Alternatively, the sensing function can actually be incorporated at a molecular level. This is achieved by combining a binding site and a reporter group in one molecule (Scheme 7.5). The reporter group is chosen to have electrochemical or spectroscopic properties that are altered by the proximate host-guest interaction. This electrochemical or spectroscopic output can therefore be used to quantitatively detect specific guests.

Electrochemical sensors

Electrochemical sensors can be created by attachment of a redox-active group to a receptor. For such a sensor to be useful, the receptor should be *selective* for the guest of interest and the binding process must be *coupled* to the redox reaction; in other words the redox active centre must 'feel the presence' of the bound guest. Many redox-active groups have been incorporated into this type of molecular sensor; e.g. ferrocene, quinone and bipyridinium (Fig. 7.4). So far, the coupling has been realized through one or a combination of the following four pathways:

(a) Through-space electrostatic interaction between the redox centre(s) and the complexed guest molecule (**7.5**);

(b) Through-bond electrostatic communication: typically provided by a conjugated linkage between the redox centre(s) and the binding site (**7.6**);

(c) Direct coordinate bond formation between the redox centre and the complexed guest (**7.7**);

(d) Induced conformational perturbation of the redox centre(s) caused by guest complexation (**7.8**).

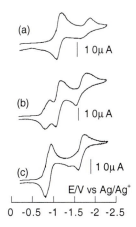

Scheme 7.5 A molecular sensor containing a reporter group (**R**) senses and reports the presence of a guest species (**G**).

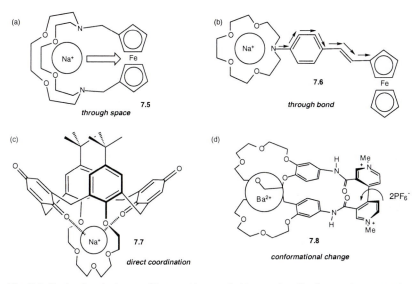

Fig. 7.4 Electrochemical recognition must be coupled to complexation for a redox-sensor to work. This coupling may be (a) through space, (b) through bond, (c) direct coordination to the redox centre or (d) be triggered by a conformational change on complexation.

Fig. 7.5 Cyclic voltammograms of **7.7** in acetonitrile in the absence (a) and the presence of 0.3 equiv. (b) and 1.0 equiv. (c) of sodium cations. New reduction waves corresponding to the sodium complex are seen to evolve as Na^+ is added (b). In the presence of 1.0 equiv. Na^+ only redox processes corresponding to the complex are seen (c) (reprinted with permission from Inorg. Chem., 1997, 36, 5880, Copyright1997 American Chemical Society).

A change in the redox properties of the receptor can be detected by an electrochemical technique such as cyclic voltammetry (CV). Figure 7.5 shows how the cyclic voltammogram of receptor **7.7** alters in the presence of

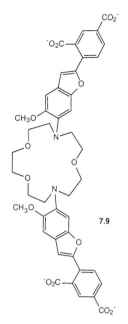

Fig. 7.6 Fluorescent probe for Na$^+$

sodium cations. Changes in the cyclic voltammogram can therefore be used to sense the presence of this guest. This approach to sensing has also been extensively used for anionic guests (for example receptors **3.23-3.25**).

Optical sensors

The commonest type of optical sensor is *fluorescent*, combining a binding site with a fluorescent unit (fluorophore). If there is efficient coupling between the two, the sensor detects the presence of the bound guest through its fluorescent output. Fluorescent sensors are especially attractive as they give a meaningful physical output which is easy to measure even at very low concentrations (light can be detected in very small quantities). They are, therefore, very *sensitive* and suitable for use in biological systems.

In spite of these advantages, the number of commercial fluorescent sensors on the market is still relatively small. Sensor **7.9** is commercially used to monitor physiological levels of sodium ions (Fig. 7.6). The crown ether binds the guest cation, the charge on which alters the electric field experienced by the fluorophore. This changes the wavelengths and intensities of fluorescence (absorption/emission) and allows the concentration of sodium ions to be determined. The sensor is popular with biologists as it is easily calibrated and can pass through cell membranes (when esterified).

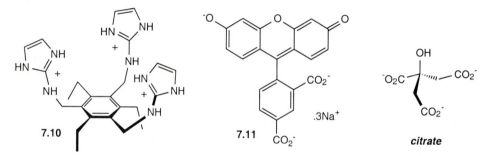

Fig. 7.7 Selective receptor for citrate anions (**7.10**). **7.11** binds weakly and allows the system to be used as a fluorescent sensor.

Chemists are now developing sensors for more structurally demanding guests in competitive solvents; a challenging goal. An excellent example is provided by a sensor for citrate anions developed in research from the group of Eric Anslyn. Tridentate guanidinium based receptor **7.10** shows a high affinity and selectivity for the tricarboxylate citrate anion (Fig. 7.7). Neither of these two components, however, are inherently fluorescent, and in order to convert the receptor into a sensor, a clever strategy was utilized. A mixture of **7.10** and carboxyfluorescein (**7.11**) was made. Substrate **7.11**, which is fluorescent, binds to the receptor, but quite weakly, as it only possesses two carboxylate groups. When tricarboxylate citrate is added to the mixture, it therefore displaces **7.11**. The fluorescent properties of **7.11** change considerably on its release from the complex, and in this way, a sensory response to the addition of citrate is obtained. Model solutions showed the sensor to be very accurate and soft drinks (including Orange Juice, Coca Cola and Mountain Dew) were then investigated. The sensor detected the concentration of citrate anions even in the presence of common anionic contaminants present in beverages, such as ascorbate and phosphate.

7.4 Switches and Molecular Machinery

As computer technology becomes ever more complex and miniaturized, scientists have begun to see the limitations of existing technologies for progressing to even smaller, more efficient devices. It has therefore been proposed that instead of using silicon chips, the next wave of technology could use individual molecules for controlling and storing information, acting as on-off *switches* and *logic gates*. This approach to '*machines on the nanometre scale*' (molecule-sized machines) has generated much excitement, and there are now many examples of molecules with controllable and addressable physical properties. In particular, the development of devices which use light as the input or output are of interest, as light has maximal velocity and can be easily controlled using fibreoptics.

A.P. de Silva and co-workers have developed a series of molecular switches and logic gates. Compound **7.12**, for example, acts as a switch. When no protons are present, the anthracene is not fluorescent, as photoinduced electron transfer (PET) from the amine nitrogen to the aromatic rings, quenching the fluorescent emission. In the presence of protons, however, the nitrogen atom protonates, preventing PET from occurring, and emission from the anthracene unit is therefore observed (Fig. 7.8). *Protons switch on the fluorescence.* This molecule could be considered as a molecular logic gate, because dependent on proton concentration there are two possible responses: on and off (1 and 0).

Using similar principles, an AND logic gate was developed (**7.13**). This receptor possesses two distinct binding sites: an amine group (binds protons) and a crown ether (binds sodium ions). In the absence of either guest PET can occur from both of these binding sites to the anthracene unit, and fluorescence is quenched. Only when sodium binds to the crown *and* a proton to the amine are *both* PET pathways inhibited; the molecule becoming fluorescent (Fig. 7.9). The action of this type of switch is reversed by adding free base and cryptands to scavenge the guest ions from the switch.

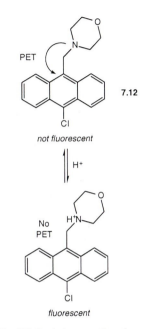

7.12

not fluorescent

H^+

No PET

fluorescent

Fig. 7.8 Basis for operation of a molecular 'on-off' switch.

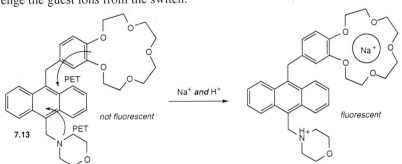

Na^+ *and* H^+

not fluorescent

fluorescent

7.13

Fig. 7.9 A molecular AND switch: fluorescence is only turned on when both protons *and* sodium cations are present.

Physically mobile 'molecular machines' have also been made. For example, the orientation of interlocked molecules such as rotaxanes and catenanes (Chapter Six) can be controlled and their spatial organisation used to store information. As yet, unfortunately, there are no actual commercial devices based on these elegant principles, but it seems certain that molecular electronic devices will soon function on a truly nanometre scale.

7.5 Supramolecular catalysts

This book has particularly focused on molecular recognition. In biology, however, binding is often a prelude to catalytic conversion of the bound substrate, and supramolecular chemists have also made a number of approaches to achieving the goal of specific, controlled catalytic conversion.

Orienting reactive and labile groups

The guest may be bound to a receptor in such a way that *reactive groups are positioned close to its labile bond*. Receptor **7.14**, developed in Jean-Marie Lehn's research group, illustrates how this approach can lead to bond cleavage. The crown ether binds the ammonium cation of the guest, and this orients the guest so that the nucleophilic thiol can attack its labile ester carbonyl group releasing *para*-nitrophenol. Interestingly, the receptor shows some of the features of an enzyme. It is substrate selective and enhances the reaction rate with slow but definite catalytic turnover (*i.e* the catalyst is regenerated). As a consequence of its chirality, **7.14** is enantio-selective for suitable optically active substrates. Its operation can also be inhibited by the addition of alkali metal cations which block the crown ether binding site.

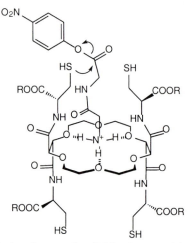

Fig. 7.10 Chiral substituted crown ether (**7.14**) acts as a catalyst for ester cleavage.

In a similar manner, functionalized cyclodextrins (see Chapter 5) have frequently been used for catalysis. The cyclodextrin cavity binds the guest and a functional attachment on one rim of the cavity then reacts with the substrate.

This general approach of orienting reactive and labile groups is analogous to that used in the design of self-replicating systems (see Chapter Six).

Raising the effective substrate concentration

Julius Rebek and co-workers took one of their self-assembled capsules (similar to **6.61** only larger) and performed a Diels Alder reaction inside it. This type of reaction is of particular interest, to chemists as there is no enzyme which can catalyse it. In the absence of the capsule, the Diels Alder reaction was slow (no product observed after one week), but when the capsule was added the reaction was accelerated 200-fold and the product formed within a day (Fig. 7.11). This acceleration is believed to be due to an increase in the *effective concentration of reactants*, which are localized inside the capsule. A bi-molecular reaction such as the Diels Alder is strongly dependent on reactant concentration (eqn. 7.1). Unfortunately, the product also binds within the cavity and so the free catalyst is not regenerated (no turnover).

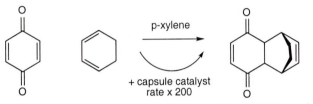

Rate = k[A][B]   (7.1)

k = rate constant

Rate law for a bimolecular reaction between A and B.

Fig. 7.11 Diels Alder reaction accelerated inside a self-assembled capsule.

Transition state stabilization

Perhaps the most familiar approach to catalysis is Pauling's postulate that the receptor can *stabilize the transition state relative to the starting materials* (by binding it more strongly) and in this way enhance the rate of the forward reaction.

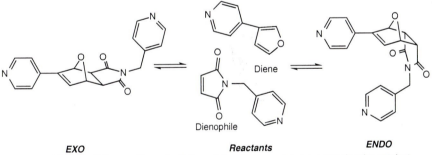

EXO   *Reactants*   *ENDO*

Fig. 7.12 Diels Alder reaction selectivity switches to exo using a supramolecular catalyst.

Sanders and co-workers used their porphyrin trimer (**6.30**) to stabilize a transition state in this way. The Diels Alder reaction they investigated (Fig. 7.12) uses pyridine functionalized substrates which can bind to the zinc atoms of **6.30**. The reaction has two different products: endo (kinetic product) and exo (thermodynamic product). Without the catalyst (**6.30**) at equilibrium the products are in a 4:1 exo: endo ratio. However in the presence of **6.30** the product is purely the exo form (no endo product could be detected *at any stage* during the reaction). The catalyst therefore stabilizes the exo transition state, probably because the pyridine nitrogen atoms are more suitably aligned for

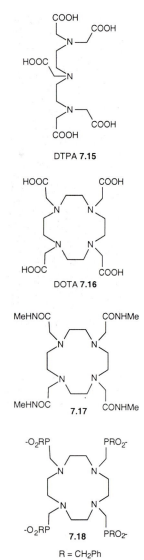

DTPA **7.15**

DOTA **7.16**

7.17

7.18

R = CH₂Ph

Fig. 7.13 The gadolinium complexes of DTPA **7.15**, DOTA **7.16**, **7.17** and **7.18** are slow to dissociate and have application as MRI contrast agents

complexation to **6.30** in this form. This illustrates the use of transition state stabilisation by a supramolecular catalyst to change the outcome of an organic reaction.

Supramolecular catalysis appeals to the innate interests of chemists wishing to use structure to control reactivity and also has great potential for application, particularly when traditionally demanding organic conversions can be efficiently catalysed, perhaps even with unusual regio- or stereo-selectivities.

7.6 Pharmaceuticals

Supramolecular chemistry is now at a stage of development where it can be successfully applied in the field of pharmaceutical research.

MRI contrast and anti-cancer agents

Figure 7.13 shows some paramagnetic lanthanide complexes which are contrast agents for magnetic resonance imaging, some of which are in current clinical use. Contrast agents are designed to accumulate in a particular part of the body (e.g. the cardiovascular system, the liver, or in tumours) and enhance visibility in MRI (magnetic resonance imaging) scans (due to the high paramagnetism of the lanthanide ion enhancing water proton relaxation) so allowing any abnormalities to be seen. An essential feature of these complexes is that they are kinetically inert and so do not release toxic lanthanide ions in the body before they are excreted.

Texaphyrin is an expanded porphyrin first synthesized by Sessler and co-workers (Fig. 7.14). The two metal complexes Gd-Tex (**7.19**) and Lu-Tex (**7.20**), shown in Figure 7.14, were found to localize in *cancerous tissue*.

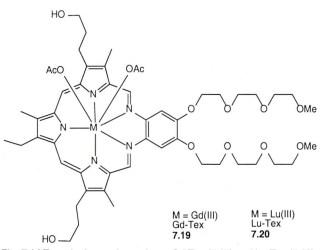

M = Gd(III)
Gd-Tex
7.19

M = Lu(III)
Lu-Tex
7.20

Fig. 7.14 Texaphyrin metal complexes Gd-Tex (**7.17**) and Lu-Tex (**7.18**).

The gadolinium complex Gd-Tex is currently undergoing trials for use as a radiation sensitizer. A cancer patient is dosed with **7.19** which accumulates in the tumour. Later the patient is exposed to radiation. It is currently thought that Gd-Tex, exists in cells as a dication. When exposed to ionizing radiation *in vivo* the Gd-Tex molecule captures an electron and becomes a π-radical cation. This process of electron capture is thought to augment the effective local concentration of hydroxyl radicals, the dominant cytotoxin produced by ionizing radiation. It is also believed that Gd-Tex and its radical cation serve to inhibit various repair enzymes including those associated with DNA repair. Figure 7.15 shows two brain scans of a patient before and after treatment with Gd-Tex. A brain tumour is clearly visible on the upper scan which vanishes after treatment with Gd-Tex and exposure to radiation (lower scan).

The lutetium complex (**7.20**) is currently showing promise as a *photodynamic therapy agent* for destroying tumours. This complex also localizes in tumour cells and when exposed to visible light, is excited to a short lived exited state that may cross over into a long lived triplet state. The triplet state can convert normal triplet oxygen into singlet oxygen (a highly reactive form of oxygen that destroys cellular components, killing the tumour). The light can de delivered in a laser beam or along a fibre optic cable and so the site of drug action is limited to the tumour itself. Texaphyrin absorbs far out to the red (at 732 nm for the Lu-Tex complex). This is an advantage over other currently available PDT agents because at these wavelengths flesh is transparent (if you look at a bright lightbulb through your hand, red light will pass through). Therefore Lu-Tex could potentially be used to treat tumours that are deeper in the body than other photosensitizer drugs. Another photodynamic therapy agent Photofrin™ (a mixture of porphyrin products) has already received approval for clinical use; the only PDT sensitizer to receive such approval so far.

Bicyclam: anti-HIV activity

Molecules of interest to supramolecular chemists can also have surprising activity against important disease targets. Linked bicyclam **7.21** (a receptor for two transition metal cations - Fig. 7.16) shows potent inhibition of the HIV virus at an early stage in its replication cycle. It is possible that this inhibition is mediated by transition metals. Synthetic chemists have tuned both the properties and activity of the potential drug by varying the spacer group.

Drug design

The pathways of disease are becoming increasingly well understood on the *molecular level*. Many enzymes which play key roles in the development of diseases, such as HIV, have had their structures elucidated. A rational approach to curing such diseases is to find a molecule that binds very strongly at the active site of such an enzyme, turning off its usual function: an *inhibitor*. Therefore, many medicinal chemists look at the structure of the active site to try to 'design' a better inhibitor. This approach requires a precise understanding of the interactions between molecules combined with considerable computer power to visualize the orientation of modified inhibitors in the binding pocket. This fundamental approach to drug design, however, can yield improved inhibitors.

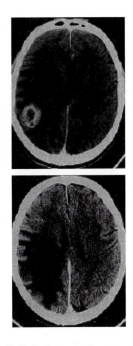

Fig. 7.15 Brain scans of a cancer patient injected with Gd-Tex before (above) and after (below) radiation treatment (reprinted with permission from gdtex.cm.utexas.edu/tex.html, Copyright 1998 Pharmacyclics Inc., Sunnyvale, CA, USA).

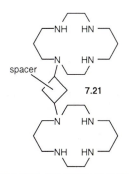

Fig. 7.16 Linked bicyclam shows potent anti- HIV activity

A schematic diagram of a new protease inhibitor bound at the active site of HIV protease is shown in Fig. 7.17. The inhibitor has specific design features: a) preorganized cyclic structure has low entropic cost of binding; b) favourable hydrophobic interactions between the inhibitor and the enzyme; c) interactions between OH groups and aspartate residues; d) carbonyl oxygen forming two hydrogen bonds with the isoleucine (Ile) residues of the enzyme.

Cyclodextrins (see Section 5.2) have found applications as drug solubilization agents. Drugs that are poorly soluble in water may form soluble inclusion complexes with cyclodextrins (the drug is bound in the hydrophobic cavity of the cyclodextrin). The drug can then be delivered in smaller doses because of its increased solubility.

Ethylated cyclodextrins release their bound guests slowly and have found application as sustained release drug delivery agents. Because the release of the bound species is slow, drug concentrations in blood can be maintained over extended periods – thus reducing the number of doses required.

Additionally drugs that are physically or chemically unstable may be shielded from the enviroment *in vivo* when included in a cyclodextrin cavity. This increases the stability of the drug therefore increasing its usefulness.

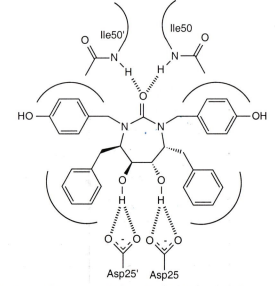

Fig. 7.17 Inhibitor for HIV protease shown schematically bound in the enzyme active site.

7.7 Summary and conclusions

Supramolecular chemistry is now a mature research area, and as such, it is surprising that there are very few books which attempt to survey the field at an introductory level. We hope that in this short text we have provided a guidebook to the principles of molecular recognition, building from simple recognition to more complex, functional systems. In this final chapter, we have tried to give an insight into the excitement that drives the field forwards, and should, in the near future, lead to an increasing number of 'supermolecules' becoming a part of everyday life.

Suggested further reading

For an excellent comprehensive survey of supramolecular technology see: Comprehensive Supramolecular Chemistry, Vol. 10 (Supramolecular Technology), Ed. J. L. Atwood, J. E. D. Davies, D. D. MacNicol, F. Vögtle, Elsevier, Oxford, 1996. For a description of molecular logic devices see: A.P. de Silva et al., *Nature*, **1993**, *364*, 42-44. For a review of artificial ... supramolecular emphasis see: A.J. Kirby, *Angew. Chem.,* **96**, *35*, 707-724. For information on texaphyrins the reader Pharmacyclics website www.pcyc.com. For information on molecular recognition and sensors see: P.D. Beer, P.A. Gale, *lv. Phys. Org. Chem.* **1998**, *31*, 1–90.